Nervous Systems

Nervous Systems

Brain Science in the
Early Cold War

ANDREAS KILLEN

HARPER

An Imprint of HarperCollins*Publishers*

HarperCollins books may be purchased for educational, business, or sales promotional use. For information, please email the Special Markets Department at SPsales@harpercollins.com.

FIRST EDITION

Designed by Leah Carlson-Stanisic

Library of Congress Cataloging-in-Publication Data has been applied for.

ISBN 978-0-06-257265-3

23 24 25 26 27 LBC 5 4 3 2 1

*To the memory of my father, and to my
mother for taking such good care of him*

Contents

PART III

Preface:
Experiments on Consciousness

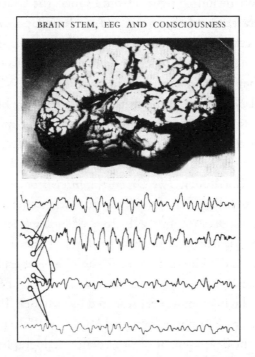

BRAIN STEM, EEG AND CONSCIOUSNESS

*During the 1950s, electroencephalography (EEG) readings
became common features of scientific and popular scientific works.
Death, formerly defined by heart failure, was redefined as brain failure,
measurable by EEG. Its scribbles became virtually a symbol of life itself.*

The resort town of Sainte-Marguerite-du-Lac-Masson, Quebec, is located in the Laurentian Mountains about an hour's drive north of Montreal. It has long been a vacation destination for the inhabitants of that city. It was in that picturesque setting, in late summer 1953, that a group of leading scientists from Europe and North America gathered to take part in a symposium devoted to the topic "Brain Mechanisms and Consciousness." During that weeklong gathering, representatives from the fields of neurophysiology, neurosurgery, psychology, and psychiatry, along with a lone psychoanalyst, met in a large chalet called the Alpine Inn to present their work and discuss the current state of a field that was then undergoing dramatic change.

Though dominated by North Americans, the group was transatlantic in composition, including figures from France, the United Kingdom, Belgium, Italy, Switzerland, Germany, and one from Japan. Among those gathered in what the published proceedings called this "inspiring mountain setting" were several in particular who are central to the story told in this book. They included the eminent neurosurgeon Wilder Penfield of the Montreal Neurological Institute, best known for his pioneering work on the treatment of patients suffering from epilepsy. Also present was his colleague the psychologist Donald O. Hebb of McGill University in Montreal, an occasional collaborator of Penfield and the author of the recently published book *The Organization of Behavior* (1949), later recognized as a founding text of modern neuroscience (claimed by some to be second in importance, in the history of biology, only to Charles Darwin's *On the Origin of Species*). From the United States came a diverse group that included the Harvard neurophysiologist Robert Morison and the Yale psychoanalyst Lawrence Kubie, another central figure in this story. The United Kingdom was represented by two men, Edgar Adrian, a 1932 Nobel laureate in physiology, and W. Grey Walter, both closely associated with the development of electroencephalography (EEG), one of the revolutionary medical technologies of the era. Through its readings of brain-wave activity, EEG had opened up new pathways for the diagnosis and treatment of epilepsy, the uncanny disease long seen as a form of "sacred possession," which was fully recognized as a disorder of the brain only in the twentieth century. Grey Walter, the last of this book's main protagonists, was the most colorful presence at the symposium. A scientific maverick based at the Burden Neurological Institute in Bristol, England, he was a neurophysiologist, cybernetician, and science fiction author and the inventor of flicker, a method of triggering epileptic seizures by means of strobe, or rhythmically flashing light. Walter's 1953 book *The Living Brain* has an important place in this story: it did more to introduce midcentury brain science to the public than any other single work, and its account of flicker went on to enjoy a strange prominence within the 1960s counterculture.

The Laurentian Symposium was dominated by discussion of the implications of recent research that, by shedding new light on the integrative action occurring in the brain, had inspired reconsideration of the relationship between brain and mind. In the words of one participant, the central question of the proceedings was to clarify how "conscious experience may be related to neuronal mechanisms in the intricate circuits of the brain." Speakers explored this question in papers on such topics as memory, learning, perception and attention, sleep and wakefulness, and illnesses ranging from epilepsy to schizophrenia.

Though it was hardly possible for the participants in the symposium to foresee the later emergence of neuroscience (a term first coined in the 1960s), they shared a belief that recent developments in their fields had brought them to the threshold of major breakthroughs. What was the source of that belief? In part it was fueled by the emergence of new techniques of studying brain activity, among them EEG, which became emblematic of an era of scientific advance that promised to open up areas long seen as off limits to inquiry. "We live," proclaimed a midcentury NIH publication dedicated to the medical applications of brain science, "in a world of marvels, of inventions undreamed of only a generation ago—rockets, space travel, machines of incredible complexity." Yet for sheer complexity, the pamphlet continued, no technological marvel could compete with the brain. In ways that had only recently come to be fully appreciated, the brain was a continually active organ: thinking, of course, but also constantly regulating vital functions such as body temperature and breathing in order to maintain the complex state known as homeostasis, without which life could not continue. Drawing from cybernetics, the interdisciplinary field whose emergence after World War II helped shape the new paradigm of the ever-active brain, the pamphlet went on to describe the complex flows of information that operated continuously throughout the nervous system, even during sleep. Much of this information flow took place below the level of consciousness, though in dreams or in the hallucinations that accompanied some neurological disorders it manifested itself in vivid forms. For this marvelous organ was not without its vulnerabilities: injury, shock, or stress, and the resulting impaired functioning, demanded precise medical intervention. The new picture of brain activity captured by techniques such as EEG meant that such intervention was now at last possible for diseases like epilepsy, the "ancient scourge of man" that had so long baffled medical science.

The uses of EEG were not confined simply to the study of illness. Participants stressed that patterns of brain-wave activity could now be correlated with mental functions such as learning, attention, and memory. According to the French neurophysiologist Alfred Fessard, psychologists had hitherto fashioned a science based purely on behavior, taking no account whatsoever of consciousness, an entity they deemed too subjective to be amenable to scientific inquiry. Consciousness was the "incommunicable" of psychic activity, best left to poets, philosophers, and mystics. But that attitude, Fessard asserted, was increasingly untenable in the new era of brain science—an era in which skilled surgeons such as Wilder Penfield were "beginning, so to speak, to 'experiment' on consciousness and raise the question of its localization in the brain." Hitherto most brain activity had taken place, as it were, offstage, scientifically speaking. But new forms of observation, measurement, and experimentation had ushered in an era of research in which the drama of consciousness itself was put onto center stage. This was, as some scholars have suggested, the beginning of the era of the "explorable brain."

But what exactly did it mean to "experiment" on consciousness? In his talk, "Studies of the Cerebral Cortex of Man," Wilder Penfield presented findings from his operations on patients with epilepsy. Over the preceding decades Penfield had developed a surgical approach known as the "Montreal procedure." Its novelty lay in the fact that his patients remained fully conscious during the procedure: the insensitivity of brain tissue to pain allowed him to operate using only a local anesthetic. Cutting away part of the skull to expose the tissue of the temporal lobe, Penfield lightly touched it with an electrode, triggering verbal responses from the patient that helped him pinpoint the area requiring surgical removal. By reporting the telltale aura that often accompanies epilepsy, patients became active participants in the procedure. But they also reported a variety of other, quite unexpected phenomena. Penfield found that they occasionally experienced visual and aural hallucinations and that in at least some cases those seemed to be related to long-forgotten scenes. Such hallucinatory apparitions, he suggested, "are re-enactments of certain experiences from the recent or distant past." By the 1950s, Penfield had shifted from seeing those phenomena simply as hallucinations to seeing them as memories. His paper about his so-called memory patients was striking for conveying not simply the new drama of midcentury brain science but its new figurative language. Drawing on the analogy of the modern mass media to illustrate his point, he observed that each reenactment is "like a strip of film from a moving picture. It is not a still picture but a moving progressive sequence of images that occupied minutes of time. These strips of film are still available to the electrode for a long time, perhaps always." In a particularly vivid formulation, he noted that patients were both actors in and directors of that cinematic production, whose images and soundtrack were "previously edited as to meaning and emotion by the patient himself."*

In his account Penfield brought together the findings of contemporary brain science and the analogical possibilities of the modern mass media to update the philosopher William James's notion of "stream of consciousness." Yet Penfield's conclusions, and the language in which he framed them, were not accepted unequivocally. Some of his colleagues expressed skepticism about his reliance on technological analogies, believing that they suggested an overly facile model of phenomena that still remained only partially understood (a view echoed much later in criticism of the computer model of the brain that also emerged during the 1950s). In years to come, central aspects of Penfield's work would be called into question, among them his notion that memories were stable, permanently archived, and always available for replay. In a strange twist

---

* Though the language is similar, Penfield's conception of the filmlike properties of consciousness differs from the contemporary neuroscientist Antonio Damasio's notion of the "movie in the brain." See Damasio, *Looking for Spinoza* (New York, Harcourt: 2003), 198, 207, 215.

of fate, a neurosurgical operation carried out less than a week after the conclusion of the Laurentian Symposium, which inadvertently destroyed the patient's capacity to form new memories, would revolutionize the study of memory. Over the course of decades of research, first initiated by Penfield and Hebb's colleague Brenda Milner, the unfortunate patient, known as H.M., would become the most famous in the history of neuroscience.

Many participants in the symposium, however, were fascinated by the hallucinations Penfield's patients reported and quick to draw parallels to related findings in their own research. The study of hallucination seemed to crack open some of the deepest mysteries of the brain. In his comment during the discussion of Penfield's paper, Grey Walter—who later spoke at length of his own experiments on consciousness in connection with his EEG studies—compared the effects of electrical stimulation of the cortex to similar effects he produced by means of stroboscopic flicker. The rhythmic flashes associated with flicker also produced hallucinatory images and memories that in some instances, he surmised, might be related to traumatic experiences. When Penfield responded by expressing doubts that the events recollected had any deeper psychiatric significance, the psychoanalyst Lawrence Kubie voiced a respectful dissent. Kubie made a plea for careful examination of "the latent emotional impact of evoked memories which may *seem* trivial." In his own later contribution to the proceedings, Kubie referred to psychoanalytic word association experiments he had conducted with Penfield's patients while they were undergoing an operation. Those experiments, he claimed, had raised the possibility that the memories activated by Penfield's electrode were related to the Freudian unconscious.

As this exchange demonstrates, there remained deep disagreement among participants about how to interpret the phenomena they observed in their research. What was not in dispute, however, was the significance attached to the fact that such phenomena were now *observable*. Whether in a neurosurgical operating theater, an EEG clinic, or a psychological laboratory, they had become amenable to measurement, analysis, and interpretation. Consciousness could be opened up to scientific study in a wholly new way—even if questions remained about whether it could be defined as a single entity. Indeed, cautioned Fessard, the best that might be said was that we are from moment to moment "differently conscious."[*]

Another illustration of the new approach would be supplied by the psychologist Donald Hebb. While expressing some doubt regarding Penfield's claims for the stability of memory, Hebb discussed his own related findings from an entirely different experimental context. He described the experiences of volunteers who had participated in an experiment in "sensory deprivation." Rather than electrical or strobic stimulation, he reported, "our method is more like a lack of stimulation." Remaining for hours or even days in a chamber designed to minimize sensory input, subjects reported hallucinations that were, he said, "quite like what is observed in mescal intoxication, as well as what Dr. Grey Walter has produced by exposure to flicker." In greater detail than the other speakers, he described the content of those visionary experiences, which tended to follow a regular course of development from simple to complex: with the eyes closed, the visual field modulated from dark to light; next there were reports of dots and lines or simple geometrical patterns and isolated objects; and finally, in a handful of cases, "integrated scenes usually containing dreamlike distortions." Subjects also reported experiences of "otherness" or "strangeness": one, for instance, reported that there were "two of me"; another the feeling that his head had become detached from his body. Such experiences, he went on, were familiar symptoms of certain forms of migraine, referring to the newly identified "Alice in Wonderland syndrome" and reminding his audience that the author Lewis Carroll had suffered from migraine. It was left to Penfield to compare the phenomena seen in sensory deprivation with those he had observed, whether produced by electrical stimulation or epileptic discharge in the temporal lobe; Kubie, meanwhile, connected them to the results of experiments with hypnosis.

Midcentury brain research, as these exchanges illustrate, explored

---

[*]   Christof Koch has stressed the major role played by epileptic patients in consciousness research, a field he describes as being still seen as "fringy" in the 1980s. Christof Koch, "The Footprints of Consciousness," *Scientific American Mind* 28, no. 2 (March 2017): 57.

many strange pathways. Like Alice going down the rabbit hole or through the looking glass, experiments with electrical stimulation of the brain, flicker, sensory deprivation, hypnosis, and word association opened portals to new realms. In so doing they mimicked the properties of the epileptic seizure itelf, which the late-nineteenth-century British neurologist John Hughlings Jackson had called a "natural experiment" for its capacity to reveal new states of consciousness. Whether hallucinations, memories, or distortions of body image or the sense of time, those states shared several distinct characteristics. As was suggested by Robert Morison, a neurophysiologist recently appointed to a position at the Rockefeller Foundation, they seemed to point to evidence of decoupling or dissociation of different brain systems, a failure of the integrative action that was a hallmark of consciousness. In the general discussion that concluded the proceedings, it was Morison who linked these phenomena to one of the strangest portals of all—namely, "brainwashing." Taking his cue from Hebb's description of the hallucinations produced by sensory deprivation, Morison proposed the following interpretation: "Hypnotism or brainwashing could . . . be thought of as a situation in which the hypnotist or brainwasher becomes able to substitute his choice of external sensory patterns for those which should be arriving over the subject's own sensory system but aren't, because of the dissociative state previously induced." Brainwashing, too, was in this sense an experiment on consciousness.

Morison here referred to a phenomenon that had recently emerged as one of great and immediate political consequence. Press reports covering the Korean War, which in the weeks immediately following the symposium ended in an armistice, had dwelt upon the uncanny behavior of American POWs, some of whom seemed to have fallen under the spell of a form of "mind control" during their time in captivity. Throughout the 1950s, this phenomenon occupied a central place in official and public discourse about the Cold War. As one government report put it, "It is man's consciousness with which the Communists experimented on American POWs in North Korea." But it also held a significant place within the contemporary sciences of brain and behavior, whose collective expertise was mobilized in response to the specter of brainwashing. Within a few years of the Laurentian Symposium, several of its participants, including Donald Hebb, Grey Walter, and Lawrence Kubie, took part in another symposium held in Boston in 1958 on the topic of sensory deprivation—a topic that had, in the interim, become central to investigation into the experiences of POWs. Though the proceedings were chiefly devoted to the new picture of brain activity that had emerged over the preceding decade, Hebb's opening remarks to the symposium dealt frankly with the political context surrounding his original experiments in sensory deprivation: "The work we have done at McGill University began, actually, with the problem of brainwashing." In some of those experiments, as he reported elsewhere, he had coupled reduced sensory input with selective exposure to propaganda and found that the combination seemed to increase subjects' suggestibility to a measurable degree.

Morison's passing remarks draw our attention to the wider historical context in which this symposium took place. Brain science was not alone in undergoing major changes in the 1950s. It was a transformative decade in many other ways as well. It is now often remembered as a halcyon, almost mythic time, a time of growing material plenty when, according to some later commentators, the American Dream reached its zenith. But the halcyon fifties were also a decade of profound social churn and apprehension about the consequences of that churn. If events in far-off Korea were one source, developments closer to home also aroused unease. Expressions of such unease were often voiced in relation, for instance, to the growing impact of the mass media on society. There as well, it seemed to some observers, "experiments in consciousness" were being carried out. It was in the 1950s, wrote the historian David Halberstam, that the nation became wired for TV, a new medium—no doubt thinking of Wisconsin senator Joseph McCarthy—"experimented with by various politicians and social groups." Similarly, when the sociologist David Riesman wrote about the forces remaking American society in the postwar era, he stressed the impact of radio, television, and movies. Subjected to the constant "bombardment" of those media, the modern individual, he wrote in The Lonely Crowd (1950), was becoming increasingly "other-directed," imprinted by the values of an emergent mass consumption society, a condition he contrasted unfavorably to an older ideal of the more autonomous "inner-directed" personality.* That older ideal was jeopardized in a social climate in which "relations with the outer world and with oneself are mediated by the flow of mass communication." The cinematic metaphors used by Wilder Penfield attest to the way in which, during that era, the mass media penetrated into the very language of scientific discovery.

The new sciences of brain and behavior advanced bold claims about the transformations sweeping American society. Harold Wolff, a neurologist and brainwashing expert based at New York Hospital, observed in the midfifties that the changes occurring in an age of rapid technological advance had brought about a "disturbance of man's total relation to his environment." In what could serve as an epigram for the era, he noted that the "human personality is not as stable as we often assume." A similar point was made by Robert Morison following a meeting in Montreal with Donald Hebb a few months after the Laurentian Symposium, when he wrote that "the importance of Hebb's sensory deprivation experiments is hard to exaggerate. It suggests that intactness of personality is critically dependent on constant feedback from the environment." He mused that this finding "makes one want to ask David Riesman whether there can be any such thing as an 'inner-directed' personality." At the end of the 1950s, Hebb himself asked whether character or personality was "an entity that exists so to speak in its own right, no matter where or in what circumstances . . . ?" He went on to observe that "In the Korean war the Chinese Communists gave us a shocking answer: in the form of brainwashing. The answer is No."

Another important voice in this story was that of Norbert Wiener, the cybernetician whose theories loomed large over the Laurentian Symposium. According to Wiener, the changes occurring at midcentury far surpassed those of the Industrial Revolution: the latter had replaced muscle power, but developments in the realm of computing and

---

* Riesman's comments were echoed by the POW Frank H. Schwable, who compared his brainwashing in Korea to the sensory bombardment of American advertising. See Elie Abel, "Schwable tells of POW Ordeal," *New York Times*, March 12, 1954.

automation, including the advent of the so-called electronic brain, fore-told the eventual devaluation of brain power, with major implications for the future of work. Hannah Arendt, the political philosopher whose book *The Origins of Totalitarianism* played a key role in shaping Western attitudes during the Cold War, reflected on some of the implications of this shift in a little-known essay on cybernetics. There she wrote of both the new possibilities and the threats tied to these developments. Foremost among them stood the question of memory: though it could be erased in computers, its erasure in people, she wrote, constituted a form of "brain-washing." Memory was central to many of the scientific investigations of the era, and the brain's entry into public consciousness during that time was often tied to unease about that fact. If human personality was not as stable as formerly assumed, perhaps that discovery was above all true of memory. Midcentury popular culture wove numerous variations on the theme of the amnesiac subject, notably in the iconic political thriller *The Manchurian Candidate* (1962).

ROBERT MORISON'S PRESENCE AT THE symposium also draws our atten-tion to two other important narrative threads of this story about the brain at midcentury. These concern the new funding climate of the post–World War II era and the new prestige of the life sciences, a prestige greatly enhanced by the discovery in 1953 of the helical structure of DNA. In a memo Morison wrote at the time, he noted that most of the participants in the Laurentian Symposium, which he praised as the best of its kind he had ever attended, had received Rockefeller Foundation funding within the previous three to four years. But the foundation was simply one of many funding sources for researchers in the field. Broad changes in the climate within which scientific research was carried out occurred in that era. Driven by the enormous demands of the Second World War, science had become organized and funded on a large scale, much of it by the fed-eral government. The most dramatic result was the Manhattan Project, the immense scientific and bureaucratic undertaking that culminated in the dropping of atom bombs on Hiroshima and Nagasaki. Though that project confirmed the status of physics as the queen of sciences, the new climate that spawned it had spillover effects elsewhere. If physics was the principal beneficiary, the new institutional arrangements forged in the 1950s spread to a host of other disciplines, galvanizing large-scale research in the life sciences: biology, medicine, psychiatry, and allied fields; those developments laid the groundwork for their eventual displacement of physics as the dominant discipline by the end of the century.

Though much research in brain science was oriented to the problems of peacetime society, at the dawn of the Cold War it was also called upon to address the new forms of unconventional warfare associated with that conflict. As numerous authors have shown in tracing continuities between the torture scandal that erupted in the early 2000s and mid-twentieth-century research into "brainwashing," the Manhattan Project was succeeded in the early 1950s by what the scholar Alfred McCoy has called a "Manhattan Project of the Mind." That project cemented an alliance between the national security state and the sciences of brain and behavior whose effects, as that scandal shows, remain with us today.

Even as the scientists gathered at the Laurentian Symposium relaxed in their mountain idyll, world events intruded into their discussions, forcing them to reckon with the fact that the object at the center of their discussions—the brain—remained an uncanny, shape-shifting entity. Brain science might now have the capacity to treat ancient scourges like epilepsy, but it also had to contend with the strange new afflictions and vulnerabilities produced by the conjuncture created by the "age of extremes" and the acceleration of social change at home. Fifties America, often remembered as an optimistic, relatively placid society, was in many ways anything but. If it was a land of material plenty, it was also a home front, a society continually imagined to be under threat by hidden forces that sought to assert control over people's inner lives. Whether external or internal, that threat was often analyzed in terms that drew upon the findings of contemporary brain science. Specialists in that field were in turn profoundly affected by the hopes and professional opportunities but also by the deep anxieties that accompanied the first decade of the Cold War. The proceedings of the Laurentian Symposium repeatedly bear witness to the entry of what the historian Richard Hofstadter called the midcentury "paranoid style" not just into the realm of politics but into that of scientific discovery as well. In what follows we will encounter numerous expressions of this.

THE STORY OF THE BRAIN at midcentury cannot be told without reference to two figures who loom large over twentieth-century human science: Sigmund Freud and Ivan Pavlov. At first glance, it is true, neither seems centrally related to this story. Though the 1950s were a golden age for psychoanalysis, its relations with contemporary brain science remained uneasy at best. Freud had begun his career as a neurologist yet abandoned that field following his discovery of the unconscious, a psychic entity with no known relationship to the brain. Wilder Penfield, for one, though prepared to discuss consciousness and its mechanisms at the Laurentian Symposium, voiced his doubts about the unconscious by asking Kubie to identify the neural mechanisms responsible for it. Though conceding that that entity remained hidden for the moment behind what he called an "iron curtain," Kubie expressed hope that new techniques might be able to penetrate behind that veil. Midcentury brain scientists found it much easier to make room for Pavlov than Freud in their theories. Pavlov's school of experimental physiology, which in preceding decades had become the "master discipline" for study of the human organism, enjoyed its own considerable prominence in the 1950s, even if it was often mistakenly associated with a reductive strain of behaviorism that took no account of brain or consciousness—a misunderstanding that, as we shall see, was often overlaid with elements of the paranoid style. Though American behaviorists tended to disregard consciousness, Pavlov, as Kubie further pointed out, had gone from banning the use of the word in his laboratory to accepting its usage. And as we shall later see in connection with Kubie, though psychoanalysis and Pavlovian physiology are often seen as quite distinct, even hostile, there were surprising convergences between the two in the 1950s.

A FULL SENSE OF THE interconnections traced above has remained largely peripheral to the origin story of modern neuroscience. To take one

example: when the neurologist Oliver Sacks, a leading voice in the telling of this story, published an article in the 1980s titled "The Prisoner's
Cinema: Sensory Deprivation," he referred to Hebb's original research
demonstrating that the brain compensated for the absence of normal sensory input by producing visionary images, hallucinations, and dreams.
Sacks echoed Hebb in drawing parallels between those findings and
those made in relation to epilepsy and migraine by Penfield and others. He did so, however, without taking note of the Cold War context
of Hebb's findings. But that context mattered a great deal, and it is for
this reason that we cannot treat the 1950s simply as a prelude to the rise
of modern neuroscience in the 1980s and 1990s. The "Decade of the
Brain" announced by President George H. W. Bush on July 17, 1990—a
massively funded program of investment into brain research and public
education—followed closely on the heels of the end of the Cold War,
during a period quite different from the scientific, political, and cultural
constellation that marked the beginning of that half-century-long conflict. As we shall see, the leap from neurological syndrome to political
syndrome—and sometimes back again—later captured by Sacks in the
notion of the "prisoner's cinema" was one repeatedly made by scientists
during that era. This is one of many reasons the decade needs to be taken
on its own terms and not treated simply as a prelude to later developments.

Though the brain's late-twentieth-century emergence as, to quote the
scholar Elaine Showalter, "the rising star of the body parts" now has an
air of inevitability, it was far from obvious that it was destined for that
status. For much of recorded history, the heart was considered the most
vital of the body's organs, "the seat of reason, emotion, valor and mind."
Contemporary accounts of the developments that turned the brain scan
into an icon of modern science surpassing the atom tend to focus on the
1980s and 1990s as the moment when new imaging technologies such as
functional magnetic resonance imaging (fMRI) ushered in modern brain
science. The result has been a somewhat ahistorical view of this development. Only recently has this begun to change, with the belated recognition that the 1950s constitute a major turning point in the field's history,
the moment in which the brain was, in Penfield's words, anointed as the
"master organ." Yet more remains to be understood about the factors that

contributed to making that era such a turning point. This book tells the story of the period of intense scientific ferment that occurred at the dawn of the Cold War, a period in which the brain was studied, experimented on, and imagined in a host of new ways—many, as we will see, intimately related to the geopolitical drama that formed its backdrop; many related to concerns closer to home.

A recent account of the transformations in brain science during the 1950s has called it the most revolutionary decade in the history of the field, one in which virtually all the foundations for the field's future were laid. Did the participants in the Laurentian Symposium see themselves as revolutionaries, as participants in what the historian of science Thomas Kuhn called, at the beginning of the 1960s, a paradigm shift? It seems unlikely. As one of the speakers noted, although the symposium might well represent a milestone of sorts, future investigators would undoubtedly look back on the "groping efforts of the mid–20th century" with "head-shaking sympathy." By the same token, however, contemporary neuroscience's understanding of its relation to its own past, particularly to the entanglements among science, politics, and culture that are part of that past, remains partial at best. It is those entanglements that will be emphasized in the pages that follow.

This book traces the stories of several of the major figures who gathered in August 1953 at the Alpine Inn and their contributions to an emergent new paradigm of brain science: Wilder Penfield, who pioneered techniques for touching the brain with an electrode; W. Grey Walter, who taught the world to read and alter brain waves by means of EEG and flicker; Donald O. Hebb, who forged new ties between psychology and biology and who invented the field of sensory deprivation; and Lawrence Kubie, who, through his experiments with Penfield's patients, sought to unite brain science and psychoanalysis. Their supporting cast includes figures such as Brenda Milner, Norbert Wiener, Robert Morison, Harold Wolff, and John C. Lilly. Their work also spawned, as we shall see, many mutant offspring, both political and cultural. Using pioneering techniques for studying the brain, its capacities and its vulnerabilities, midcentury scientists opened up pathways to the future even while that future remained clouded in uncertainty. If the revolution that occurred in the 1950s had a founding manifesto, it is Grey Walter's *The Living Brain* that holds the strongest claim to that status, and it is to that text that we turn in chapter 1. That will be followed by the first of a series of clinical tales whose inclusion in this book is intended to make the perspective, and where possible the voice of the midcentury patient, part of the story that follows.

PART I

CHAPTER 1

Brain Science at the Dawn of the 1950s

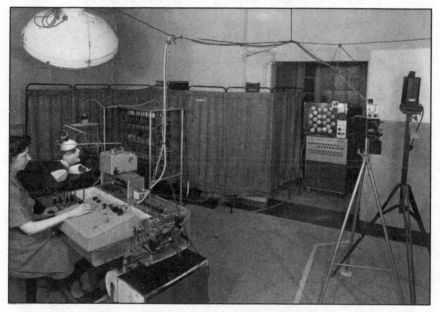

The Toposcope Laboratory.

W. GREY WALTER, *THE LIVING BRAIN* (LONDON:
GERALD DUCKWORTH AND CO., 1953).

The scene in the grainy photograph bears the caption "A moving panorama of the brain rhythms." It depicts the layout of a novel space: the Toposcope Laboratory, a chamber in which EEG readouts of patients' brain waves are recorded while they are subjected to the effects produced by rhythmically flickering light. On the image's left-hand side, a man reclines on a couch. Suspended just above his head is a stroboscopic lamp that generates a flicker effect, while next to him sits a woman at a desk, operating the control panel of a recording system. Against the wall to the right-hand side of the image, a monitor displays the brain-wave readings in the form of projected images.

The image appears in the Anglo-American neurophysiologist Grey Walter's book *The Living Brain* (1953), and the space it portrays is central to that book's account of contemporary research in brain science. As Walter explained, the strobe lamp generated a range of light spectra and flicker effects that produced sharply varying responses in the subject's brain-wave patterns. The resulting EEG readings helped identify anomalies in brain function, including those associated with epilepsy, thus enabling clinicians to zero in on areas associated with the seizure disorder. But, as Walter also explained, flicker could sometimes induce epileptic auras in healthy subjects. Those experimentally produced seizure phenomena were occasionally accompanied by remarkable hallucinations as well as unusual emotions and even memories.

The technique described by Walter is emblematic of an array of mid-century methods of capturing what went on inside people's heads. As the photo conveys, those methods commonly involved elaborate technical arrangements: complex laboratory or clinical apparatus, sensitive recording devices, and methods of analyzing sizable quantities of data. Such ensembles spoke of the 1950s as a new age of technoscience that heralded the transformation of many fields of medicine. Almost unnoticed in this photograph, though, is one element that evokes a rather different approach to getting inside people's heads, the couch, which recalls a quite different setting: that of the psychoanalytic office. Its presence here hints at an alternative paradigm of human science, reminding us of the fact that, even as figures such as Grey Walter announced the dawn of a new era in brain science, the 1950s were also the golden age of psychoanalysis in America and Great Britain, the moment when the teachings of Sigmund Freud advanced from the margins to the center of society. As will become clear in the chapters that follow, the two approaches to getting inside people's heads had very different objects in mind: for one it was the brain; for the other the mind. Yet the boundary between the two did not always remain stable in the work of scientists of that era.

This midcentury drama of the brain and mind was complete with a large and colorful cast of protagonists. In the late 1940s and early 1950s, a diverse crowd of specialists emerged on the scene, announcing startling discoveries in neurology, psychiatry, behavioral science, and allied fields. They included figures such as the eminent neurosurgeon Wilder Penfield, whose operations on patients with epilepsy helped unlock some of the brain's deepest secrets; the cybernetician Norbert Wiener, whose name was closely linked with the first accounts of the "electronic brain" (or computer); Donald O. Hebb, whose experiments helped forge new links between psychology and neurophysiology; the psychologist Brenda Milner, whose pioneering studies of memory helped lay the groundwork for the later emergence of the field of cognitive neuroscience; and Walter Freeman, the United States' foremost practitioner of lobotomy (or psychosurgery). They also included a new generation of psychoanalysts, whose contributions to treating the wartime problem of battle neurosis greatly enhanced their authority in postwar America; a generation of antipsychiatrists critical of the invasive methods of figures such as Freeman; and new forms of healing such as Dianetics (or Scientology), the creation of a former science fiction author, L. Ron Hubbard. Though the differences between the approaches those figures represented were significant, taken together, their appearance at that moment speaks to the fact that the 1950s were characterized as few periods previously by the desire to get inside people's heads—an aspiration that, as we shall see, was accompanied by much social and political drama.

A central figure in this story was Grey Walter. Though his name is not well known today, he was at the center of many of the transformations enumerated above. Born in the United States, he spent most of his life in England, holding a position at the Burden Neurological Institute in Bristol until his retirement in the 1970s. His career was built on his contributions to the rapidly developing field of EEG, which came into its own in the decade after World War II. It was also shaped by his association with the cybernetics movement, which his close friend and ally

Norbert Wiener had first introduced to the public in the late 1940s. In particular, Walter gained public stature and no small amount of notoriety through his invention of a kind of mechanical brain or automaton, the so-called tortoise (so named after one of the creatures in *Alice's Adventures in Wonderland*). Walter emerged as a self-appointed spokesman for the revolution in brain science that occurred in the aftermath of World War II, and his book *The Living Brain* served in many ways as that revolution's manifesto. It became a scientific best seller in the United Kingdom, the United States, and beyond, introducing the reading public to the remarkable developments then occurring in the field.

The Prehistory of Modern Brain Science

The story that Grey Walter unfolded for his readers was epic in its scope, full of bold claims, speculative leaps, and surprising twists. Before embarking on his tour of the new panoramas opened up by contemporary research, he first took pains to establish the novelty of the brain's emergence as a new frontier and proper object of scientific study. That was a surprisingly recent development. For millennia, Walter wrote, the brain's place within the hierarchy of the body's organs and its relation to higher functions had remained a matter of dispute. Indeed, he wrote, for many of the ancients it had been of secondary importance to other organs such as the heart. The Greeks themselves had had no word for it, referring to the gray substance contained within the skull as *encephalos* (inside the head). They tended to associate the mind or soul with regions below the neck such as the heart or stomach, a tendency preserved in the etymology of modern terms such as *schizophrenia*, which derives from the Greek term *phren* (diaphragm). The philosopher Plato associated dreams and hallucinations with the liver; much later, the medievals assigned human temperaments, or "humors," to organs such as the gallbladder (choleric), the blood (sanguine), and so forth. The result of these and other such theories, according to Walter, was that "the brain remained in the dark for over 2000 years."

This is not to say that the brain had not been an object of fascination over the centuries. Philosophers, theologians, and scientists had speculated about its relation to thought, intelligence, morality, emotion, and other essential features of human identity. Yet well into the nineteenth century, uncertainty persisted. The philosopher William James identified the viscera as the seat of the emotions, while many doctors continued to associate the classic "female malady," hysteria, with the womb. Freud may have helped to shift its locus to the nervous system, yet he eventually abandoned his early neurophysiological studies for an idiosyncratic topography of the mind that had no relation to the brain.

The brain's emergence as *primus inter pares* among the body's organs was thus, in Walter's telling, a belated development. Paradoxically, however, recognition that the brain was the seat of all that was human placed it off limits to scientific research. Seen as the most refractory of the body's organs, its sheer complexity made it daunting to even the most intrepid researchers, many of whom also harbored deep reservations about extending their experimental methods beyond the top of the spinal cord to the mind, emotions, morality, and creativity. As Walter wrote, those taboos had remained in place at Cambridge, where he had carried out his postgraduate work in neurophysiology under the Nobel laureate Edgar Adrian in the early 1930s. Meanwhile, the two dominant schools of early-twentieth-century psychology, behaviorism and psychoanalysis, ignored the brain entirely.

In the Beginning Was Feedback

In Walter's account, the modern "heroic" age of brain science emerged out of the confluence of two landmark developments: the Russian physiologist Ivan Pavlov's work on conditioned reflexes and the German psychiatrist Hans Berger's invention of electroencephalography (EEG). It was Pavlov, through his arduous, decades-long study of reflex conditioning—a process centrally coordinated within the brain—who broke the taboo against the study of that organ. And it was Berger who in 1928 supplied a method of directly investigating brain-wave activity. The method employed electrodes attached to the scalp to pick up the minute signals emitted by the basic wave patterns: beta (the brain in a state of arousal), alpha (the brain's normal resting state), theta (linked with daydreaming), and delta (the sleeping brain). By means of an apparatus that moved pens across continuously scrolling paper, those signals were transcribed into scribbled lines. Using those wave patterns and anomalies in them to draw conclusions about brain function and dysfunction, scientists could now study the brain as a living organ rather than as an object of postmortem investigation. They could also begin to relate neural mechanisms to memory and learning—hitherto the province of those who studied the mind (a term about whose empirical basis Walter made no secret of his skepticism).

Those advances coincided with a significant shift in attitudes toward the brain. The sources of this shift were many and complex, but Walter singled out several as especially notable. Among them was cybernetics, the interdisciplinary field born during World War II that made a significant contribution to the midcentury study of the human nervous system, brain, and mind. Cybernetics arose out of the convergence among communications engineering, mathematics, neurology, psychiatry, and anthropology, which came together around the study of how systems— technological, biological, social—regulate themselves to maintain stability. The theories of information and communication central to that movement played a formative role in helping pry open what had hitherto been treated as the "black box" of the brain and of mental processes. Cybernetics also contributed to the conceptualization and design of artificial or "electronic" brains. The first reports of such devices appeared in the press in late 1948 and early 1949, and their appearance marked a significant element in the theory and practice, as well as the public image, of brain science. Cutting-edge work in the field derived much of its glamour from cybernetics, which was promoted as a charismatic new master discipline based on the foundational concept of feedback: the property, as Wiener put it, of organisms and machines to modify their future performance on the basis of past performance. In the beginning, wrote Grey Walter, was feedback: thus his book announced its theoretical debt to cybernetics in its first pages with a typically provocative flourish by invoking feedback as "the first act of creation."

Further evidence of this paradigm shift was the advent, in the fields of neurology and psychiatry, of an array of new, often invasive forms of treatment, among them neurosurgery, electroshock therapy (ECT), and lobotomy. The widespread adoption of those new treatments signaled the radical change taking place in attitudes to the brain. Though Walter, himself a pioneer of ECT in Great Britain, did not dwell at length on those sometimes controversial methods, what little he did say was largely positive. Their hallmark was a new boldness, a willingness to intervene directly in the tissues of the cortex. Not least among the contributions of neuro- and psychosurgery was their role in advancing the mapping of the brain and catalyzing research on hitherto little-known structures within it. As the Russian neuropsychologist A. R. Luria later wrote, until recently "Precise knowledge [of the brain] was rarely to be found in the textbooks . . . maps of the brain [were] scarcely more reliable than medieval geographers' maps of the world." The middle of the twentieth century was a crucial moment in the process of filling in the blanks in the uncharted territory.

It is hardly surprising that those breakthroughs were accompanied by a sense of unease. For his part, Grey Walter had no doubts about their significance in bringing about, as he put it, "man's tardy recognition of the organ that alone sets him apart from other living things." Yet even as he identified the brain as the seat of man's distinctive essence, his book was full of disturbing references to the human brain's likeness to a variety of nonhuman—whether mechanical or animal—brains. In the public realm, where these developments received wide coverage, their materialist implications were met with fascination but also dismay, even shock. Reporting on the Nobel Prize awarded to Egas Moniz, the Portuguese inventor of lobotomy, in 1949, the New York Times noted that his technique seemed to have rendered the brain "no more sacred than the liver." The fact that the living brain was now treated as simply another organ to be scrutinized by the medical gaze may well have been a scientific breakthrough of the highest importance, yet it was laden with philosophical, moral, and epistemological significance whose implications were far from self-evident.

Some sense of that ambivalence is captured in French critic Roland Barthes's short essay on Albert Einstein's brain, which became an object of public fascination following his death in 1955. For Barthes, the great physicist's cerebral organ was a quintessentially mythical object, in the modern, twentieth-century sense of that term: "Paradoxically, the greatest intelligence of all provides an image of the most up-to-date machine, the man who is too powerful is removed from psychology, and introduced into a world of robots," where he joined the "supermen of science-fiction." The celebration of Einstein's brain as a "prodigious organ" coincided with that organ's loss of all "magical" properties; transformed into purely physiological substance, a great ethical chasm was opened up, leaving Barthes to wonder what remained of the scientist's conscience.

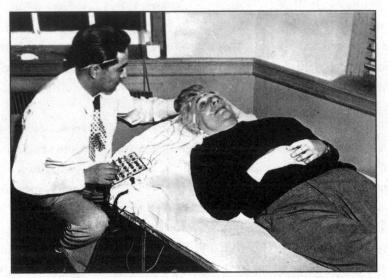

Einstein hooked up to an EEG: "thinking of relativity."

NEW YORK TIMES (1951). AP-PHOTO/NAP.

Barthes went on to invoke a photograph taken while Einstein was still alive. It showed the great scientist hooked up to an EEG. Lying down, his head bristling with electric wires, he was asked to think of relativity. "What this is meant to convey is probably that the seismograms will be all the more violent since 'relativity' is an arduous subject." Thought became in that way something measurable, quasi-electrical, the product of a material organ made "monstrous only by its cybernetic complication." Einstein himself was elevated to the status of a genius, though one whose thought performed a kind of labor analogous to the mechanical making of sausages. Death, Barthes wrote, was above all, for him, the cessation of a localized function: *"the most powerful brain of all has stopped thinking."* Yet despite the reduction of Einstein's brain to a purely material organ, a pure calculating machine, the equations that it produced nevertheless had a decidedly esoteric quality: in $E = mc^2$ the formula for nature and the universe, he mused, was revealed in Gnostic fashion.

Know Thy Brain, Know Thyself

Though deeply grounded in the science of its time, the tone of Walter's book was literary, sly, provocative. The transgressive qualities noted by Barthes in connection with the new way of thinking about the brain permeated his prose. Those qualities in turn mirrored the colorful persona that Walter adopted for himself. Born in Missouri, he spent most of his early years living abroad before being sent for his education to England, where he remained the rest of his life. Something of an outsider within the conservative culture of Great Britain's scientific establishment, he cultivated an identity as iconoclast and breaker of taboos. By the time his book was published in 1953, when he was forty-three years old, he had begun to pursue a wide range of extrascientific interests that included TV punditry, science fiction authorship, wife swapping, and drug experimentation, as well as left-wing, even anarchist politics (the latter sympathies being part of a family tradition he inherited from his father). Yet for all his nonconformist attributes, he also took evident delight in surrounding himself with the material trappings of the automated, mass consumption society that gradually emerged in postwar Great Britain. In a late essay he recounted in loving detail the features of his domestic idyll: a detached house with garden, two cars, motor scooter, five bedrooms, color TV, central heating—a catalogue adding up to what he called "the image and symbol of normal practical creativity and convenience." So automated was the Walter household that for a time it also included, as we shall see, two mechanical pets.

But if Walter's love of gadgetry extended deeply into the fabric of home life, it found expression above all in the grand claims he made for EEG. He regarded the method as the key to many fundamental questions of existence extending far beyond the diagnosis of epilepsy—a technique that would crack open the secrets of the brain. That view was shared by many of his closest allies, including Norbert Wiener. Having been introduced to EEG by Walter, Wiener believed that its readings confirmed the brain's essentially cybernetic nature and identified the hieroglyphic scribblings of brain activity as the "Rosetta Stone" of cognition and intelligence. Walter's own heady sense of EEG's importance was coupled with a belief in the necessity of establishing strict control over the standards governing its use. He saw himself as an empire builder in the emergent field of EEG, its "pope," in the words of one observer. He sought to position it as the cornerstone not simply of clinical practice but, more ambitiously, of the management of entire societies. Some sense of the grandiosity of that vision found its way into the remarks of contemporary observers, among them representatives of the Rockefeller Foundation, to which Walter submitted frequent funding requests. Testimonials to his brilliance were accompanied by references to his growing sense of self-importance and the faint aura of the mad scientist that he cultivated. The Rockefeller Foundation's Robert Morison, for one, was deeply impressed by his mastery of the arcana of the EEG but noted with distaste the "singularly Mephisphelean beard" that he began to sport. That hint of the outré in Walter's persona also found its way into accounts of his clinic: following a tour of the Burden, another visitor compared Walter's Toposcope Laboratory to "a nightmare TV or movie set, with a jungle of cables and everywhere the exposed bowels of various pieces of electrical apparatus." The comparison was not inapt, given, as we shall see, the many parallels Walter drew between brain activity and the mass media.

Yet if his personal style, and perhaps his politics, kept him from scaling the rarefied heights of Great Britain's scientific establishment, Walter compensated by cultivating a public profile that made him a much sought after and widely read figure. Knowledge of one's brainwave patterns, he boldly proclaimed, was the new key to self-knowledge. Those patterns were defining features of personal identity. He noted, for instance, that strong alpha rhythms correlated with weakness of interior imagery and proudly identified his own small alpha waves as the neurophysiological correlate of the great gifts of internal visualization that had helped make him a scientific pioneer (as one observer put it, "Grey Walter had no alpha wave at all until after the second cocktail"). That essential new form of self-knowledge, he went on to claim, had important implications in many realms of life, not least within the global political arena. He believed that EEG readings might hold the key to both marital harmony and negotiations between the superpowers. Yet even as he indulged in such proclamations, he tried to avoid giving offense to his audience's more traditional sensibilities. In the foreword to *The Living Brain* he took pains to reassure readers worried about its "risky" subject matter, the indecency "of brain surveying brain." His, he claimed, was a "pious" book, reverent toward "Man"; it contained "no immodest exposure, no baring of the soul." After finishing his book, many readers may well have wondered, however, what place if any was left for the "soul" in its materialist account.

Walter was a gifted popularizer of scientific knowledge. In lively prose punctuated by repeated references to Lewis Carroll and William Shakespeare, his first chapter sketched the natural history of the brain from the earliest stirrings of life on Earth. With feedback as "the first act of creation," some primitive species began to evolve the organized nervous systems that conferred advantage over their rivals. An epic process of trial and error spanning millennia laid down the basis of neural evolution, a process whose intermediate stages and many wrong turns remained largely a matter of guesswork: "Nor shall we ever know," he mused, "what innumerable freaks and monsters, their mutations being expendable, fell back unseen into the abyss." Brains emerged first among fish, then among birds and reptiles. At each stage of the process, learned behaviors imprinted themselves in the neural organization, giving rise to increasing levels of complexity. A major watershed was reached when mammals developed internal temperature control. That, wrote Walter, "was a supreme event in neural, indeed in all natural history." Not only did it enable survival on a cooling planet, but it created the basis for an automatic system for stabilizing the organism's vital functions. This was the steady state known as homeostasis: an *active* state maintained through constant feedback loops to preserve stability in the body's internal environment amid changing external conditions. The emergence of that function enabled other, higher functions to develop. Thus the brain benefited from an increasingly complex division of labor, allowing different regions to pursue specializations such as observation, memory, and judgment. Yet although the process of evolution and differentiation opened up ever new horizons, the brain preserved within itself remnants of "old landmarks": "The experience of homeostasis, the perfect mechanical calm which it allows the brain, has been known for two or three thousand years under various appellations." Walter identified that mechanical calm as the tantalizing ideal sought after by masters of yoga and other such practices.

Having reached that point in his account, Walter then dispensed in a few pages with what he regarded as the largely negligible achievements of the prehistory of modern brain science before turning to the two giants whose work, in his view, marked the real beginnings of the field: Ivan Pavlov and Hans Berger. It was the confluence of their efforts in the first half of the twentieth century that set the field on the right path. With them the study of the brain and its functions passed from the prescientific past into the scientific age. Those beginnings, however, were not simply to be located in some pure moment of laboratory creation. With a dramatic flair that would have done credit to a writer of speculative fiction (a genre he later dabbled in), Walter's creation story was inextricably bound up with war. At the outset of the Second World War, he observed, only two EEG laboratories had existed in the United Kingdom; by the end, there were fifty. The war had produced untold numbers of brain-related injuries, the diagnosis and treatment of which had necessitated a radically new approach. At the first meeting of the National EEG Society, "held literally in the midst of the Battle of London, criteria of electrical abnormality [the kind associated with brain trauma] were agreed on." The agreement reached at that meeting was followed by a veritable flood of EEG readings. Since then, "Literally millions of yards of paper have been covered with frantic scribblings." Walter treated that development as nothing short of epochal; it created grounds for hope, he claimed, that the EEG would "provide the mirror which the brain requires to see itself steadily and whole." Immense new vistas had thereby been opened up.

Dr. Walter's Cabinet of Wonders

What were those vistas? Let us return to the photograph with which this chapter opened, which establishes one of the principal settings in which midcentury brain science unfolded. In the hands of Grey Walter and its other pioneers, EEG became the key to the process of detecting, recording, and interpreting the electrical wave patterns that accompany brain activity. Electrodes attached to the scalp registered the minute voltages continually discharged by the brain and transmitted the signals to the recording apparatus. To the untrained observer the signals were bewildering, but with practice the main patterns of brain activity could be studied and differentiated into the basic wave types, ranging in speed and amplitude from highest to lowest states of arousal. Identifying spikes or other anomalies in wave patterns enabled the clinician to assess and localize the severity of brain affliction. In particular, Walter stressed the role such readings played in registering the electrical storms that accompanied epilepsy. Variations in the oscillating pattern of brain waves could be correlated with seizure activity, and by that means the epileptogenic area could be located in preparation for surgical intervention.

The process of collecting and interpreting the signal data emanating from the brain demanded highly sensitive equipment. Walter's chief contribution to that task was the so-called toposcope, a multichannel recording system capable of capturing a much higher degree of information than an ordinary EEG. Walter compared its relationship to the original technique to that of a "mosaic of aerial photographs . . . to a traveler's tale." By obtaining simultaneous readings from a multitude of cortical areas, the machine, which employed cathode ray tubes like those used in TVs to translate the brain's electrical impulses into images that could be projected onto a screen, functioned as a kind of "brain television" that produced a composite picture or landscape, a "moving picture" of brain activity. Not for the last time in this story, we encounter a parallel between brain science and the modern mass media. This was reinforced by Walter's reliance on analogy to illustrate the division of labor relating to visual processes in the brain: "The neural situation may be compared with that of a film studio," in which the director positions cameras but leaves technical details such as focus to individual senses.

Further vistas were opened up by introducing new variables or, as he put it, "imposing other patterns on the brain through the senses." Chief among the means of doing so was strobe, or as Walter preferred to call it, "flicker." Flicker, he wrote, was the key that opened many doorways to brain function. In a semidarkened room he invited his patient to lie down on a couch and, once she was comfortable, trained a stroboscopic lamp on her eyes. The lamp produced rhythmically flashing lights at rates that Walter could adjust to specific brain-wave frequencies, the most common being alpha (in the eight-to-thirteen-cycles-per-second range). The flashing lights acted directly on the brain in a way that seemed to break down the barriers between different areas of the cortex, a phenomenon that produced effects resembling those of the electrical storms characterizing an epileptic seizure. The combination of strobe lamp and adjustable flicker rate thus made it possible to reproduce experimentally the symptoms associated with epilepsy. Valuable new data were thereby obtained. A condition formerly shrouded in feelings of "religious awe" and "horror" was now made observable and measurable. The influence of Pavlov's form of experimental physiology, the "master discipline" for the study of the human organism, reveals itself here. Just as Pavlov, in a famous set of experiments, had created "experimental neuroses" in his dogs, so, too, Walter created a kind of experimental epilepsy in his patients.

Not content with limiting his research to patients, Walter also trained the strobe lamp on healthy subjects. In a small percentage of the trials, subjects exposed to certain frequencies of flicker experienced clinically diagnostic symptoms of epilepsy: strange feelings, auras, convulsions, even, in rare cases, unconsciousness. Such laboratory results, he noted, mirrored those sometimes experienced in everyday life: the flickering light that a passenger in a car sees as he passes along an avenue of trees; or the moviegoer who, in a state of trance induced by the projector's flicker rate, tried to strangle the person in the seat next to him. Those findings confirmed Walter's sense of the revelatory possibilities of flicker. In using this method to study epilepsy, he and his team unveiled a variety of strange effects: hallucinations, emotions, anomalies in personality types. Subjects exposed to flicker reported seeing vivid and marvelous imagery: moving patterns such as whirlpools and whirling spirals, effects he compared to the results obtained by Wilder Penfield when he touched his patients' brains with an electrode (see chapter 2). Such hallucinations, moreover, were often accompanied by strong emotions—dizziness, fear, disgust, anger, pleasure—and in some cases powerful memories. Some subjects reported feeling that they had moved "sideways in time" (an experience that Walter later tried to capture in fictional form in a novel about time travel). In addition, subjects' responses seemed to be related to some extent to personality or will. Subjects hallucinated as they watched the rapid flashes, usually by "seeing" colorful figures in motion. The often cinematic qualities of the imagery were underscored by Walter's reliance on cinematic metaphors in his explanation of the mechanisms involved.

"We Called Him Tortoise"

Among the mutant offspring spawned by midcentury brain science were the automata beloved by members of the cybernetics movement. Walter's claim that EEG would "fashion a mirror for the brain" was echoed in the claims he made for such automata. Having established to his own satisfaction that electricity was the key to brain activity, it was but a short step to the construction of electromechanical devices capable of mimicking the basic elements of that activity. These were his so-called tortoises (dubbed *Machina speculatrix* in recognition of the exploratory or speculative behavior he identified as the most basic propensity of animal life). Housed in a plastic shell, the appealingly toylike contraptions were equipped with

touch and light sensors and capable of adjusting their movements in response to information conveyed by the sensors. They were first introduced
to the reading public in May 1950 in the pages of *Scientific American* in
an article titled "An Imitation of Life," prefaced by a passage from a text
much beloved by Walter, Lewis Carroll's *Alice's Adventures in Wonderland*:
"We called him tortoise because he taught us!" Over the first half of the
1950s, Walter made repeated public demonstrations of Elsie and Elmer,
as he called his first creations, most notably at the Festival of Britain held
outside London in 1951 to commemorate the one hundredth anniversary of
the landmark Crystal Palace Exhibition of 1851 (which Lewis Carroll had
himself attended). They were reputedly one of the most popular attractions
at the festival, though their appearance was marked by minor drama, having initially been rejected by the organizers because, as Walter put it, the
Pavlovian theory of conditioned reflexes on which their design was based
was deemed "un-British." The mathematician Alan Turing was among the
many fascinated by the "feedback dance" the tortoises performed in a mirror Walter included in his exhibit.

Although those public demonstrations enabled Walter to indulge his

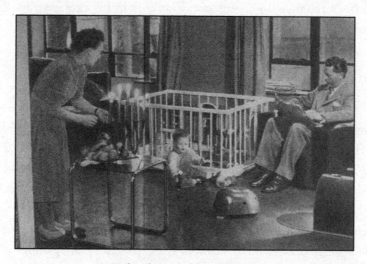

Walter household with tortoise.

flair for showmanship, the machines had a serious purpose, signaled in Carroll's association between the tortoise and learning. Elsie and Elmer were deceptively simple contraptions that used sensors and photoelectric cells to perform complex, goal-directed behavior.*

In later refinements of the basic design, Walter added—to the original two sense organs and two electronic nerve cells (all that was needed, he suggested, to demonstrate "free will")—memory circuits allowing for the storage of associations. By thus demonstrating the tortoise's ability to learn simple behaviors, he ventured into terrain hitherto treated as the preserve of those who studied mind. To do so, he summoned the authority of Pavlov, whose dogs had been trained to develop precisely these same types of associational responses or, as the Russian chose to call them, conditioned reflexes. Close investigation of those processes had also led Pavlov to develop a theory of types to explain the influence of personality on learning.

But the study of learning also opened up other fields, as well as a range of other, more troubling possibilities: the relationship not only to personality type but also to breakdowns of the learning process, excesses, or pathologies—in short, a series of properly psychiatric phenomena requiring appropriate intervention. Once a learning circuit was provided, the possibility of a conflict neurosis appeared: "In difficult situations the creature sulks or becomes wildly agitated and can be cured only by rest or shock—the two favorite stratagems of the psychiatrist." In *The Living Brain* Walter elaborated on that cryptic remark. There, in the course of describing how his mechanical tortoise could be taught to respond to a whistle by training it to associate the sound with a light detectable by its sensor, Walter touched on the results of experiments with cross-conditioning, in which conflicting incentives and penalties created confusion and agitation in the creature. Essentially he was driving his tortoises

---

* The creation of feedback loops in which environment was an element of the learning process resulted in behavior that suggested self-consciousness: "Creatures of sorcery peering into the dim 'electro-biological' future in search of a *deus ex machina* . . . look up at us and declare that God is a physiologist." W. Grey Walter, *The Living Brain* (London: Gerald Duckworth, 1953), 130–32.

mad, as Pavlov had done with his dogs. In such cases there were several ways of aiding recovery. If the conflicting associations did not die away, "Switching off all the circuits and switching on again clears all lines and provides, as it were, a new deal for all hands. Very often it has been necessary to disconnect a circuit altogether—to simplify the whole arrangement. Psychiatrists also resort to these stratagems—sleep, shock, and surgery." Conceding that many people found such stratagems abhorrent, Walter nevertheless insisted that direct attack "may well and truly arrest the accumulation of self-sustaining antagonism and 'raze out the written troubles of the brain.'"

Walter's claims for the humanoid properties of his tortoises were greeted with both fascination and skepticism. Among those intrigued by their implications was Lawrence Kubie, a psychoanalyst with a background in neurophysiology and a keen interest in cybernetics. Walter and Kubie had forged a warm friendship at the Laurentian Symposium (see preface). Kubie was a close reader of *The Living Brain*, and the correspondence between the two men is marked by many tokens of his appreciation for the book. Not the least of its virtues in his eyes was its role, together with the work of Donald Hebb on sensory deprivation, in laying to rest the notion of the central nervous system (CNS) as a passive network akin to a telephone exchange. It was one of the axioms of the new paradigm that the CNS was now recognized as a continually active system. Yet in his role as respondent to a talk Walter gave in New York in the late 1950s, Kubie raised a series of pointed questions about the limits of the machine model of consciousness. Could Walter, he asked, make a machine with a capacity for fantasy? Could he explain the unconscious? Such questions reflect Kubie's desire to find ways of bridging the divide between neurophysiology and psychoanalysis. One surprising development to come out of that was Walter's apparent conversion to this program. In a letter to Kubie recalling a mountain hike they had taken during a break in the Laurentian Symposium, Walter stated, "Your writing encourages me regularly to continue with the scheme to marry neurophysiology and psychoanalysis on earth as it is in heaven." Kubie reciprocated in kind, mentioning the work of the young neurophysiologist John Lilly (inspired by Hebb's research into sensory deprivation) as being particularly promising in this regard (see chapter 7).

Though the depth of Walter's commitment to this program remains hard to gauge, there is no question that his book was marked by a fascination with unusual states of consciousness, a fascination it shared with psychoanalysts of Kubie's ilk. From epileptic auras to flicker-induced hallucinations, from hypnosis to yoga, Walter explored many esoteric aspects of brain activity. In the case of hypnosis, for instance, he reported the fact that EEG failed to register anything out of the ordinary when confronted with subjects in a hypnotic trance. No change in wave patterns was detectable even though, under the spell of the hypnotist's suggestions, this state magnified the "power to learn" to a remarkable degree (subjects could spontaneously develop blisters, cramps, etc.). "In hypnosis," he suggested, "we see how wide and deep is the dominion of the brain over all other organs and functions." But in that state the rules of conditioning seemed to be waived. The hypnotist evidently gained access to the inner workings of the learning mechanism without affecting the basic properties of brain function.

Or consider his comments on the place of involuntary functions within the economy of brain and body. It will be recalled that Walter identified homeostasis as a landmark event in neural evolution; the fact that humans do not have to consciously manage breathing or body temperature allows them to devote some part of their mental energies to more complex functions. That was how things should be. Yet, he observed, things were not always so simple. In the case of psychosomatic disorders, involuntary processes (which, he wrote, normally remain submerged in "what we may still call the unconscious") took on a life of their own: they control the organism, preoccupying and dominating it to the point of sickness. But by the same token, he went on, we can learn to control those processes. Here, too, flicker showed the way. His studies showed that the subject's responses to its effects could be to some extent learned: "The will of the subject can be brought into play: he can, for instance, consciously and with effect resist or give way to the emotions or hallucinations engendered by the flicker"—a finding with important implications for the subject of "self-discipline." Indeed, he noted, some non-Western cultures went to fantastical lengths to develop techniques for doing so. Conscious control of involuntary functions "is the basis of grotesque cults . . . in which long years are spent in practicing a system of conditioned reflexes whereby pulse rate, breathing, digestion, sexual function, metabolism, kidney function and the like are brought under conscious control." Such feats could be studied in the laboratory as measurable conditioned reflexes. Walter's later work along such lines helped inspire the principles of biofeedback, in which individuals are trained to acquire control of precisely these functions.

Among the many fantasies awakened by EEG, as Walter also noted, was the possibility that it might provide confirmation for the existence of paranormal phenomena such as telepathy. Though he himself rejected such possibilities, they formed an essential part of the cultural reception of EEG. The notion that EEG might transcribe not simply brain waves but actual thoughts provided the premise of the German émigré and horror novelist Curt Siodmak's *Donovan's Brain* (1942), a novel adapted for film several times. This novel, which marks the entry of EEG into popular culture, tells the story of a wealthy tycoon who dies in a plane crash and whose brain is kept alive in a vat, while a scientist attempts to establish communication with the organ via EEG; eventually the brain develops telepathic control over the scientist and uses him to carry out a series of crimes. The scenario of the brain in a vat became common in midcentury science fiction and later in philosophy.

In fact, the uncanny possibilities associated with EEG were part of the very genesis of the apparatus. Hans Berger, the German psychiatrist who invented EEG in the late 1920s, was a longtime researcher in parapsychology, with a particular interest in telepathy, who hoped that his new technique would help establish the material basis for such forms of communication or mind reading. It is unclear whether Walter himself was aware of that aspect of the origin story of the method that his book celebrated, yet even as he rejected the possibility of telepathy, he remained fully alive to the esoteric associations it awakened. Those associations haunted the public reception of EEG.*

Getting Inside the Head

Strange as the associations awakened by EEG were, the procedure itself was physically noninvasive: it translated electrical activity into pictures of the brain without intervening in the tissues of that organ. This was not the case with lobotomy. Although Walter touched only briefly on that procedure, it was not by chance that the radical change in attitude toward the brain that he heralded coincided with its heyday. Pioneered in the 1930s by the Portuguese Egas Moniz, it was enthusiastically adopted in the 1940s by the American psychiatrist Walter Freeman, who promoted its benefits for patients of all types—not just those suffering from severe psychotic disorders but neurotic patients as well. Estimates are that more than forty thousand lobotomies were performed in the United States before the method fell victim to both the advent of psychopharmacological means of treatment for mental illness and the mounting revulsion against its effects.

The revolution announced in Walter's book was closely connected to what he called "the heroic physical treatment" of mental disorder. He likened psychosurgery to the process of disconnecting a circuit while hastening to add that even a lobotomized brain retained its plasticity and resilience. Yet though he spoke of astonishing cures in patients who had been "mad as hatters for years," he conceded that little was known about how methods such as psychosurgery or ECT worked or what the actual

---

* In Thomas Pynchon's *Gravity's Rainbow* similar possibilities inform the proceedings at the White Visitation, a mental hospital located on the British coast near Bristol (the site of Walter's clinic), which serves as the site of experiments in World War II–era psychological warfare. Clairvoyants and telepaths gather there and scientists study their EEG readings, looking for "unusual spike patterns" that might confirm the possibility that information could be transmitted on "strange frequencies." They come "from as far away as the institute in Bristol to gape at . . . the freaks of Psi Section." See Pynchon, *Gravity's Rainbow* (New York: Viking, 1973), 145–46.

long-term therapeutic benefit to the patient might be. The benefits to the researcher, on the other hand, were not inconsiderable. The abundance of subjects with carefully planned and reasonably uniform brain injuries, Walter wrote, presented valuable opportunities to study the reestablishment of functional patterns; "otherwise the casualties of road and battle were his only material."

Yet the views he expressed masked a certain defensiveness. The year his book was published, 1953, marked the beginning of lobotomy's decline; the appearance on the market that year of a new class of psychopharmaceuticals including chlorpromazine led to the rapid discontinuation of the practice. Some of Walter's allies within the cybernetics movement, including Norbert Wiener, had themselves long harbored deep misgivings about lobotomy. In a story titled "The Brain," published in 1952 in a science fiction magazine, Wiener signaled his profound unease over the invasive practices of modern psychosurgery (for more, see chapter 6).* That unease also found expression in what came to be known as the antipsychiatry movement.

The Future of the Brain

What place, if any, did Walter leave for mind in his account of the mechanisms and functions of the brain? Though he preferred to avoid that question, that became harder to do as he ventured into more speculative terrain, extrapolating from the findings of brain science to their implications for society and for the future. Though prepared to acknowledge the role of "personality," he saw it as an entity distinct from mind, rooted in behaviorist models or in Pavlov's neohumoral model of types. Up until quite recently the brain had been a black box, understood only in terms of inputs and outputs; theorists of education invented a content for that box—"mind"—in order to make sense of the process of learning: "Freed by

---

* Lobotomy was also critiqued in another contemporary work of science fiction, Bernard Wolfe's *Limbo* (New York: Ace Books, 1952). Influenced by Wiener, *Limbo* tells the story of a society of the future in which brain surgery is a commonly performed procedure meant to eradicate the roots of violence in man.

homeostasis for intellectual pursuits, the supreme abstraction of the brain was indeed the mind." But, Walter noted, recent developments heralded significant change: "From the confusion of metaphysics and psychoanalysis . . . the thinking brain has turned eagerly to the first possible glimpses of itself. The millennial process of its unconscious evolution ends before the mirror; a new phase begins." It was now possible to make conscious what had hitherto been largely unconscious; to think scientifically about thinking. Yet though he acknowledged the usefulness of conventions such as mind, he made clear his impatience with them. Indeed, he suggested that it might eventually be better to discard the term altogether.

> The physiologist, viewing in his modest workshop the inexplicable electrical tides that sweep through the living brain, knows that the bobbing of his float must mean some Leviathan is yet uncaught; some great idea nibbles his bait and slides darkly beneath the laughing waves. But he would be happier not to dub it Mind; he would prefer to call the one that got away—Mentality, thinking of it only as a relation of dimensions, in the same class as Velocity.

From Walter's perspective, mind was best conceived less as an entity in itself than as a particular modality of brain function.*

Into the Cold War Maelstrom

Walter's writings repeatedly underscore his indebtedness to cybernetics. World War II had been the crucible of this movement. Many of

---

* Such views were echoed in J. Z. Young's *Doubt and Certainty in Science: A Biologist's Reflections on the Brain* (London: Galaxy, 1951). Young stressed the claims of brain science over questions long claimed by other fields of inquiry. What philosophers liked to call the will, he suggested, was simply the set of rules learned from experience and stored in the cortex. It was time for them to recognize that the brain stood at the center of human activity, even during creative acts such as painting. As he paraphrased the public attitude: "'It's not true that the brain paints the picture—it is *I* who do that. I am not just a mass of whitish stuff inside my skull'" (73–75, 78). His book was devoted to demolishing the fallacy of the view that the brain was "just a mass of whitish stuff."

its leading members, including Wiener and Walter, had been involved in radar systems research during the war, research that had a significant impact on the postwar life sciences. Its principles of information and feedback, which were integral to the design of human-machine air defense systems, shaped the formation of influential new paradigms of communication and control that spread into other research areas. Cybernetics now claimed for itself the status of master discipline, a kind of unified field theory encompassing most scientific fields (though its actual importance for brain science remains a matter of some debate). Apart from its immediate practical implications in areas ranging from rocketry to communications engineering to physiology and psychiatry, it lent itself to highly speculative, even futuristic modes of thinking. Some of its acolytes were tempted to see in it the promise of a comprehensive basis for the management of society. A conference held at the Burden Institute in 1949 to discuss Wiener's ideas conveys some of the flavor of that promise in its title: "Meeting on Control Mechanisms in Machines, Animals, Communities." The meeting brought together physiologists, engineers, and political theorists to analyze the convergence of different branches of science around control mechanisms and communication.

Already in his early writings, however, Norbert Wiener had signaled misgivings about the direction being taken by the field so closely linked with his name. Those misgivings only grew as the high-speed computational systems ("electric brains") that emerged in the late 1940s were put to work on the calculations needed for the hydrogen bomb, first tested in 1953. As scholars have written, the unleashing of the atom bomb had already triggered a profound personal crisis in Wiener, who increasingly stressed the extraordinary ambivalence of cybernetics. That ambivalence was brought sharply into focus by the new superpower rivalry. In a Cold War context marked by an ever-escalating quest for new forms of technological advantage, Wiener noted that the destructive potentials of that quest tended at their furthest extreme to erase conventional friend/foe distinctions: "The enemy may be Russia at the present moment, but it is even more the reflection of ourselves in a mirage. To defend ourselves against this phantom, we must look to new scientific measures, each more terrible than the last. There is no end to this vast apocalyptic spiral." Having reached that gloomy conclusion, he ventured the hope that cybernetics might be reoriented away from war to fields such as physiology and psychology. Yet he expressed grave doubts regarding the attitudes prevalent in much contemporary psychiatry, such as those expressed by the practice of lobotomy. And in what appeared to be an anticipation of anxieties about "mind control," Wiener wrote that information, the very lifeblood of cybernetic theory, could become fatally caught up in this quest for geopolitical advantage and the culture of secrecy that was its inevitable desideratum.* The desire to get inside people's heads could, he foresaw, acquire a fateful dynamic, one that might conceivably end with cybernetics' absorption into the dark arts of modern intelligence gathering. Many leaders of contemporary science, indeed, were "nothing more than apprentice sorcerers, fascinated with the incantation which starts a devilment that they are totally unable to stop." Alluding to another aspect of the emerging mass consumer society that inspired misgivings, he concluded on a pessimistic note, "Even the new psychology of advertising and salesmanship becomes . . . a way for obliterating the conscientious scruples of the working scientists, and for destroying such inhibitions as they may have against rowing into this maelstrom."

For his part, Grey Walter remained relatively unfazed by such concerns. The darkening shadows of the Cold War seemed not to trouble him. If the foreword to the first edition of *The Living Brain* (1953) included reassuring bromides about the reverence with which the book treated "Man," the second edition of his book, in 1963, dealt in equally summary fashion with the ethical questions that had since emerged. It went without saying, he wrote, that the physiologist must oppose all destructive tendencies of the sciences, and "so far brain physiology has never lent itself to evil deeds as physics has done—we do not have

---

* Wiener's mounting mistrust of that culture eventually led him to withdraw from government work. By the 1960s, cybernetics had become increasingly caught up in the tensions of the Cold War. It had acquired a "red tinge" and become closely associated with communism. See Loren Graham, *Science and Philosophy in the Soviet Union* (New York: Alfred A. Knopf 1972), 326, 336.

to drive people insane or experience insanity to win understanding of brain action." Probably responding to Aldous Huxley's *Brave New World Revisited* (1958) (see chapter 5), he continued, "But with increasing understanding comes growing power, and there are many who already see in our tentative essays the means for effective thought-control—some with glee and some with revulsion. To all such we say, as to politicians, preachers, and psychiatrists: let only those whose hands are clean busy themselves with brain-washing." As we shall later see, the brainwashing scare that seized American society in the 1950s was, among other things, a chapter in the history of the societal reaction against the invasive practices that Walter passed over so lightly. It was articulated with particular force by members of the antipsychiatry movement, who critiqued the lobotomists' tendency to treat diseases of the mind as diseases of the brain.

In his futuristic mode Walter speculated that brain science might help resolve numerous problems of contemporary civilization. In relation to present-day youths, for instance, he suggested that anomalies in the brain's theta rhythms, which seemed to feature prominently in the EEG readings of juvenile delinquents, might hold the key to the detection and prevention of violent tendencies. Such speculations formed part of a visionary projection of a society of the future subject to the cybernetic controls of a new scientific caste. Deeper knowledge of the brain, he mused, might well help predict and control the course of future events. At the same time, he noted worrisome developmental trends in modern society. Among them were those associated with the rise of television. If society, he wrote, had seemed for a time to reflect the plasticity and adaptability of brain function, "now it seems to be degenerating into something more like a spinal cord, able to receive instructions and implement reflex coordination but incapable of initiating an independent or original idea." The symbol of that was a child transfixed by the TV screen: "A passive solitary child gazing at the screen of a television receiver amuses only itself—the need to gaze does not promote or evoke habits of creativeness or generosity." Equally troubling was the contemporary proliferation of new cults, among which he singled out L. Ron Hubbard's Dianetics, later known as Scientology. In Dianetics, he wrote with barely concealed disgust, one encountered a vulgarized form of all the contemporary practices of healing.

Yet Cold War realities also intruded into Walter's scientific idyll. Though less overtly than in Wiener's writings, the specter of totalitarianism was nevertheless a presence in Walter's book. This was manifest, for instance, in passages critical of the way that Pavlov's teachings had been distorted by the Soviets. We can predict, he wrote, that educational schemes based on misinterpretation of Pavlov are likely to produce mental derangement rather than education. Such schemes, he warned, would seek to condition a fault in the brain's conditioning mechanism: "The mechanism breaks down, sooner or later, when these natural functions are tampered with; the mind is flattened into a shallow mold; anything can mean anything and untruth be truth." Here, despite his earlier protestations, he came very close to an account of brainwashing and to the discoveries about the vulnerability of the brain that would move to the center of the new forms of ideological and psychological warfare that characterized the Cold War.

Conclusion

Perhaps it was his role as scientific popularizer that compelled Grey Walter to draw sharp boundaries between the discoveries to which he introduced his readers and the related possibilities of mind reading and mind control seemingly opened up by those discoveries. Yet even as he rejected such cultural fantasies, he courted controversy in many other respects: for instance, in suggesting that older creation stories be replaced with the new story based on the foundational principle of feedback or in suggesting that the very idea of mind was obsolete. In doing so he knowingly set himself against many reigning pieties of midcentury society. The brain that emerged from the pages of his book was, in the end, a thoroughly cultural object, deeply enmeshed in the tensions of its historical moment.

In 1970, Walter was involved in a motor scooter accident that left him with a severe brain injury. He was in a coma for three weeks and underwent a five-hour-long surgical procedure. In the hospital where he recuperated following his surgery, he was plagued by disturbing hallucinations of cyborgian surgery and prosthetic reconstruction.* By that time the future envisioned by Walter had taken on characteristics he might scarcely have imagined in the early 1950s. In an irony that surely would not have been lost on him, that was due, at least in part, to the cultural movements to which his own work served as tributary. *The Living Brain* became a central text of the sixties counterculture, which discovered in it a manual for the production of altered states such as those documented by Walter. Flicker, as we shall see, became a key to exploring ways of escaping or undoing precisely those forms of mind control that Walter so lightly disavowed but would become basic to the anxieties unleashed by the Cold War.

---

* Walter's hallucinations eerily echoed the dystopian scenario of Bernard Wolfe's *Limbo*, a novel that, partly inspired by Norbert Wiener's writings, envisioned a future society of war, lobotomy, and prosthetics.

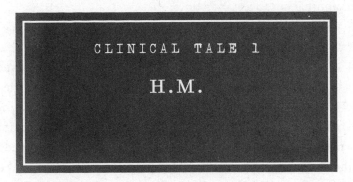

CLINICAL TALE 1

H.M.

Midcentury case histories bear witness to the frequent links between scientific discovery and scientific transgression. What eventually entered the pantheon of modern neuroscience as "the most famous brain in the world" was the result of an experimental procedure that ended in tragedy. On September 1, 1953, a young man named Henry Molaison underwent an operation on his brain in the hope of obtaining relief from his severe epileptic seizures. His physician, the Hartford Hospital neurosurgeon William Scoville, determined that radical intervention was needed and, probing more deeply than any surgeon had done before, removed part of his patient's hippocampus, most of his amygdala, and several ancillary structures making up the so-called limbic lobe. Though Molaison's seizures diminished following the surgery, the procedure also had the effect of destroying his capacity to form new memories. For the next half century up to his death in 2008, he lived in a twilight condition of what one doctor called "the permanent present tense." H.M., as he was long known to the outside world, became the most intensively studied patient in the history of neuroscience, and research on his catastrophic impairment transformed the science of memory. After his death, Molaison's brain was sectioned into 2,400 slices, each 70 microns thick (now available for online viewing as part of a project called the Brain Observatory). Contemplating Molaison's fate, one observer was moved to write, "His brain is the closest that science approaches to a saint's relic."

As a ten-year-old boy, Henry Molaison suffered a biking accident that resulted in a head injury. It was likely as a consequence of that accident that he began as a teenager to experience epileptic seizures accompanied by violent convulsions. By the time he reached his twenties, they had become increasingly frequent and debilitating, rendering normal life all but impossible. Having exhausted conventional treatments, his parents, a middle-class Connecticut couple, eventually turned to the prominent neurosurgeon William Scoville, who performed the procedure that catapulted Henry from obscurity into the annals of brain history.

To fully appreciate what made that procedure such a consequential one in the history of brain science, it is necessary to step back and consider the wider context in which the encounter between Molaison and Scoville took place. Epilepsy has for millennia been one of the most enigmatic and disturbing of disorders known to man, an ancient scourge shrouded, in the words of Grey Walter, in mystery and awe. Sigmund Freud identified it as the most uncanny of all diseases, one in which the body seems to take on a life of its own outside conscious control.* Often seen as a manifestation of demonic possession, it was also regarded as a "sacred disease" whose onset could be accompanied by visionary experience. The Russian novelist Fyodor Dostoyevsky was one of many seizure-afflicted people who described such episodes in near-ecstatic terms.

The discovery that epilepsy is a brain disorder is a relatively recent one. Up to the end of the nineteenth century, many doctors associated it with a variety of other organs (including, bizarrely, the genitalia). Those misconceptions were mirrored in the confusion surrounding

---

* One could say, one author has written, "that a wicked genie had gathered together all the imaginable symptoms together in this strange ailment." Catherine Clement, *Syncope: The Philosophy of Rapture* (Minneapolis: University of Minnesota Press, 1994), 9.

its treatment. Many physicians regarded it as incurable, and well into the twentieth century the treatment of epilepsy, like that of other severe neurological disorders of uncertain etiology, languished. The situation began to change only in the 1930s with the development of new kinds of anticonvulsant medication and new forms of surgical treatment. The latter development was pioneered by Wilder Penfield, who introduced novel techniques of neurosurgical intervention for patients suffering from epilepsy and other severe neurological afflictions (see chapter 2). To do so, he relied extensively on the expertise of his close colleague Herbert Jasper, who used EEG to help identify the brain area responsible for the affliction. EEG also helped remove another source of confusion surrounding epilepsy by establishing that it was an organic neurological condition rather than a psychiatric disorder.

Nevertheless, confusion persisted. Led by the Portuguese psychiatrist Egas Moniz and his American disciple Walter Freeman, another form of brain surgery—lobotomy, in which the connections between the prefrontal cortex and the rest of the brain are severed—was soon adopted for the treatment of psychiatric patients suffering from a spectrum of mental disorders ranging from depression to schizophrenia.[*] The majority of neurosurgeons did not treat psychiatric patients and did not perform lobotomies. Penfield himself, after briefly experimenting with lobotomy, quickly abandoned it and thereafter expressed deep aversion to the procedure, confining his operations to neurological patients. Henry Molaison's doctor, William Scoville, on the other hand, imposed far fewer limits on his use of this procedure, performing it on both psychiatric and neurological patients. Scoville was not

---

[*]   Over a twenty-year period, more than forty thousand lobotomies were performed in the United States. Though he admitted to having no precise understanding of how the procedure worked, Freeman had no doubts about its efficacy, citing the fact that highly agitated patients tended to become much calmer following the procedure, a transformation he explained as resulting from the "bleaching of affect." The method was controversial, and by the early 1950s, mounting opposition to it had developed. The discovery of antipsychotic medications led to its rapid decline. See Luke Dittrich, *Patient H.M.: A Story of Memory, Madness and Family Secrets* (New York: Random House, 2017), 165.

simply any physician but a major figure in his own right, second only to Walter Freeman in the number of lobotomies he performed as chief neurosurgeon at a hospital called the Institute for Living in Hartford, Connecticut. Well into the 1970s, Scoville remained an outlier in the profession for his insistence on the virtues of lobotomy. At 1973 congressional hearings he was attacked by antipsychiatric activists for continuing to defend the method. By that time the field of medicine had become deeply mired in ethics scandals surrounding revelations of the practices that marked much of the postwar period, even in the wake of the Nuremberg Code, which enshrined the principal of informed consent in medical practice. Psychosurgery's heyday (1945–1952) is a central chapter within a longer history stretching from the end of the war up to the early 1970s that was marked by thousands of "hazardous and invasive experiments on uninformed, un-consenting subjects"—many of them patients with epilepsy.

During the immediate postwar era, however, its enthusiasts advocated such forms of intervention as holding out genuine hope, not simply for alleviating intractable illnesses but for elucidating some of the brain's deepest mysteries. In particular, psychosurgery catalyzed scientific interest in the limbic lobe, a little understood but tantalizing group of brain structures including the hippocampus, the amygdala, and the hypothalamus that were believed to be associated with emotion, memory, and psychiatric illness. Scoville was one figure who shared that belief. Having treated Henry Molaison with antiepileptic medication for years to no avail, he decided to perform his experimental procedure on him.

Following a further series of meetings among Molaison, his parents, and Scoville, the procedure was scheduled for the end of the summer of 1953. In the waning days of August, Molaison underwent a final series of EEG and psychological tests, the latter of which revealed nothing definitive, according to the tester, beyond the somewhat curious finding of evidence of "castration anxiety" and "homosexual" tendencies. However, the EEG shed no light on the location of the lesions, and it was that negative finding that there was no clear target that convinced Scoville to adopt a more radical course of action. He opted for bilateral section. Drilling two holes into Molaison's skull, one above each eye, he used a silver straw to suction out deep-lying parts of the temporal lobes on both sides. The results of the procedure were immediately apparent: though Scoville succeeded in bringing Henry some measure of relief from his seizures, he also permanently destroyed his ability to store new memories.

As catastrophic as that outcome was, it seems to have done little to shake Scoville's sense of belief in his own judgment (though he never performed the same drastic procedure again). Four months after operating on Molaison, Scoville published, in the January 1954 issue of the *Journal of Neurosurgery*, a short article titled "The Limbic Lobe in Man," a kind of medical manifesto disguised as a modest report. In it he reported on the results of recent efforts to probe the mysteries of the limbic lobe, the deep-lying area that he had targeted in his operation on Molaison. Such procedures, he reported with evident gratification, had brought considerable relief to patients suffering schizophrenia, neurosis, and epilepsy. Apart from a general calming effect, behavioral and other psychiatric changes had been minimal, "with the one exception of a *very grave, recent memory loss*, so severe as to prevent the patient from remembering the locations of the rooms in which he lives, the names of his close associates, or even the way to toilet and urinal." Passing from that reference to Molaison to a concluding paragraph in which he allowed himself to speculate about the implications of neurosurgery for the secrets locked up in the limbic area—the fundamental mechanisms of mental illness and of epilepsy—he wrote, "Who knows but that in future years neurosurgeons may apply direct selective shock therapy to the hypothalamus, thereby relegating psychoanalysis to that scientific limbo where perhaps it belongs?" Moreover, he concluded, might not selective procedures "raise the threshold for all convulsions, and thus dispense with pharmaceutical anticonvulsants?" Those remarks—which attest to continuing uncertainty concerning the boundary between neurological and mental illness—are revealing for the way in which they place Scoville's approach within a disputatious and rapidly shifting landscape, one in which the practice of psychosurgery was shortly to enter rapid decline.

This is the point marking the entry of Henry Molaison, henceforward called H.M. to protect his privacy, into the annals of scientific legend. He became the subject of thousands of studies, and his case of total amnesia established the basis of the modern science of memory. Following the operation, Scoville contacted Wilder Penfield, who put him in touch with his assistant Brenda Milner. Milner had received her PhD under Donald Hebb, whose landmark 1949 book *The Organization of Behavior* had significantly advanced the integration of neurophysiology and psychology, two fields hitherto largely siloed from each other. Prior to that time, the brain had played little role in theories of learning, but Hebb's identification of the existence of areas of heightened synaptic connectivity (the so-called Hebbian synapse) had established the neural basis of learning.* Yet important pieces of the picture remained to be filled in, some of which could be found only by studying patients with brain damage. In reports written for the Rockefeller Foundation in the early 1950s, Hebb discussed the significance of the findings of human brain operations, noting the element of contingency they introduced into his research: "Here not much planning is possible, except to be ready to take advantage of interesting case material that happens along." As Milner put it in an autobiographical sketch, her early training had taught her to appreciate the value of studying the behavioral effects of brain lesions: insight into the functioning of a normal brain, she wrote, could come only through an "analysis of disordered function." Her affiliation with Penfield's MNI was crucial in this regard.

It was at that point that the second chapter of Henry Molaison's story—his momentous association with Brenda Milner—began. Milner

---

* Describing his discovery of the long-term potentiation of associated synapses, Hebb wrote, "Any two cells or systems of cells that are repeatedly active at the same time will tend to become 'associated' so that activity in one facilitates activity in the other." Hebb, *Organization of Behavior: A Neuropsychological Theory* (New York: John Wiley & Sons, 1949), xix. Put more colloquially, nerves that fire together wire together.

is a central figure in the history of modern neuroscience. Born in England and educated at Cambridge University under the psychologist Frederic Bartlett, a figure associated with important work on the reconstructive nature of memory, she worked in radar systems research during the war, applying her training to the selection of aircrew. She then moved to Montreal to earn her PhD under Hebb's supervision. Following that she joined the MNI to carry out psychological studies of Penfield's epilepsy patients before and after surgery (measuring, for instance, the procedure's effect on IQ)—a move, she would later write, that eventually became permanent, thus ensuring the "future of neuropsychology at the MNI."

There were few people in the world better prepared to study the psychological consequences of Scoville's operation on Molaison than Milner. She began commuting between Montreal and Hartford to conduct the first of what would eventually be thousands of hours of tests. As a result of those studies, she was able to establish both the crucial role of the hippocampus in consolidating long-term memory and the distinction between different types or systems of memory. In H.M.'s case, long-term, or narrative, memory was gone, but procedural, or task-related, memory remained intact. In 1957, Scoville and Milner collaborated on a paper based on the initial findings of their studies that became memory science's "founding text," one of the most frequently cited publications in neuroscience. For the first time, they were able to tie memory to a specific location in the brain. The studies of H.M. that Brenda Milner began in the mid-1950s (and that her successor, Suzanne Corkin, continued for close to half a century) revolutionized the study of memory and laid the foundation for wider developments. The historian Gordon Shepherd, indeed, credits Hebb, Milner, and H.M. with having together founded the field of cognitive neuroscience. Their collaboration and the transformative developments it brought about affirm the 1950s' status as a moment of central significance in the field's history. As the Russian neuropsychologist A. R. Luria, himself the author of a landmark study of memory, *The Mind of a Mnemonist*, wrote to Milner following the publication of her paper, "Memory was the sleeping beauty of the brain—and now she is awake."

And what of H.M.? He lived at home with his mother until her death, then entered a Connecticut nursing home, where he spent his remaining years before dying in 2008. By all accounts he never lost his equanimity, remaining largely oblivious to his impairment and remarkably good-natured about the tests that made up a good deal of his postoperative existence. In that regard he was an ideal experimental subject. Lacking the ability to consolidate memories, however, he became a strange presence, eternally struggling with names and faces, as well as new words. Despite their lengthy association, he never remembered Milner's name from one meeting to the next. His command of vernacular English remained frozen in the year 1953. Confronted on one test by the term *brainwash*, his answer was "The fluid that surrounds and bathes the brain." For Milner, Molaison's loss was profound: "he lost his humanness." As one author has put it, a mind without episodic memory seems fundamentally "alien"—humanly speaking, almost a contradiction in terms.

The story of Henry Molaison and his role within the history of memory science would not be complete without mention of the dispute that, following his death in 2008, erupted over custody of his brain. It had originally been intended to be housed at the University of California at San Diego's Brain Observatory, where his brain had been sectioned; the battle over the final disposition of the brain sections and the accompanying data was finally resolved in favor of the University of California at Davis, a decision reached in apparent violation of earlier agreements. Such disputes attest to the fact that whether singly or as part of collections, brains emerged in the modern era as a crucial form of scientific capital. Careers were made on the basis of access to the brains of both living and dead subjects. The disposition of Einstein's brain after his death in 1955 similarly became a matter of heated dispute. And as we shall later see, brain collections became part of the history of the Cold War in the form of the Soviet regime's Brain Pantheon (see conclusion to book).

Occasionally the disputes were about the murky questions surrounding provenance. A case in point was the controversy that erupted at the Fifth International Neurological Congress in Lisbon in 1953.* The controversy was triggered by the participation of the German neuropathologist Julius Hallervorden, whose pathbreaking research on brain defects rested in large part on a collection of more than five hundred brains that the organizers of the Nazis' T-4, or euthanasia, program had made available to him. A heated debate threatened to derail the preparations for the conference and cast a pall over the subsequent proceedings, as participants worried about the unwelcome attention that the revelations about Hallervorden might draw to their field. Such episodes illustrate the degree to which the advance of modern brain science occurred in the shadow of research of a morally fraught kind. As we shall see in subsequent chapters, questions surrounding such procedures were posed very sharply by the antipsychiatric movement that emerged in the 1950s. That movement criticized the brain's transformation into a site of experimental intervention and made the erasure of memory, both individual and public, central to its critique of contemporary brain science.

---

* The conference was notable, among other things, for the meeting that took place between the Portuguese Nobel laureate Egas Moniz and his American disciple Walter Freeman. Moniz reportedly greeted Freeman with the words "Ah my dear Freeman, you are the great expansionist of leucotomy [lobotomy]." Elliot Valenstein, *Great and Desperate Cures: The Rise and Decline of Psychosurgery and Other Radical Treatments for Mental Illness* (New York: Basic Books, 1986), 239.

CHAPTER 2

The Brain Is an
Undiscovered Country

The operating theater of Wilder Penfield.

The hospital in which Wilder Penfield performed neurosurgery was arguably the world's leading institution of brain science at mid-century. It was housed in the Montreal Neurological Institute (MNI), which Penfield had founded in the 1930s and which under his direction had grown into a renowned center of research and medicine, one in which neurosurgery, neurology, anatomy, and psychology were closely integrated. That multidisciplinarity made it a forerunner, claimed Penfield's longtime colleague Herbert Jasper, of what became known as neuroscience. The operating theater was its centerpiece. Every part of the surgical dramas that unfolded there was aided and recorded by specialized instruments. Adjacent to the theater was a built-in space for an electroencephalograph (EEG) and its operator, Jasper. Additional space was reserved for a photographer, who was tasked with visually documenting each operation. Also present was a stenographer, who transcribed Penfield's observations as well as statements made by the patient, who, as we shall see, remained conscious and verbal while undergoing surgery. Meanwhile, a spectator gallery allowed students to peer down from their tiered seats into the glass-enclosed space and observe the proceedings.

Wilder Penfield was the preeminent neurosurgeon of his time. When the son of Allen Welsh Dulles, the head of the CIA, suffered a traumatic head injury in Korea, it was to his fellow Princetonian Penfield that he turned first for advice (see Clinical Tale 3). But Penfield's primary area of expertise was epilepsy, and the theater served chiefly as the setting for his operations on patients suffering from especially intractable forms of epilepsy. Long considered incurable, this condition was surrounded by much stigma, and for many the diagnosis of epilepsy constituted a kind of death sentence. But in the 1930s, new treatments appeared that offered hope to victims of the disease. Most dramatic among them were Penfield's surgical procedures. Those bold operations brought him much renown, and by the 1950s, stories about them and the remarkable findings that he made while performing them began to appear regularly in the press.

Penfield came from modest origins. He was born in Spokane, Washington, and, following the breakup of his parents' marriage, raised by his mother in a small town in Wisconsin. Under her firm hand he gained entry to Princeton, where he excelled as both student and athlete. From there he embarked on a scientific odyssey that eventually took him to most of the world's leading centers of brain research, including

Johns Hopkins School of Medicine; Oxford University, where he studied
under Charles Sherrington; Madrid, where he learned the cell-staining
techniques of Santiago Ramón y Cajal; and Germany, where he worked
in the clinic of Otfrid Foerster (who had been called to Vladimir Lenin's
bedside following his stroke). His work grew out of his direct links to the
great European medical traditions of the early twentieth century. It was
above all under Sherrington's tutelage that Penfield learned to view the
nervous system and brain as a vast "undiscovered country." The mapping
of that terra incognita, he later wrote, revealed the human brain to be "the
master organ of the human race." This process of exploration, however,
was not without its hazards. At a time when postoperative survival rates
were low, Penfield's early career was not infrequently marked by tragedy.
His memoir includes an especially harrowing account of operating on
his own sister to remove a brain tumor; though the procedure brought
her temporary respite, she died two years after the operation. Such experiences strengthened his determination to put his profession on a more
secure scientific and institutional footing. With the help of grants from
the Rockefeller Foundation and the province of Quebec, he eventually
succeeded in establishing the MNI, or "Neuro." It subsequently became
a model for all such institutes.

One major difficulty confronting neurosurgeons lay in identifying
the area of the brain responsible for the electrical discharges associated
with seizures. EEG helped localize such areas by capturing anomalies
in brain-wave patterns. Yet it was not always able to obtain reliable
readings from deeper parts of the brain; its inability to pinpoint the
target area was one of the factors that led William Scoville to perform
the more radical operation that resulted in the destruction of H.M.'s
memory. Though Penfield relied heavily on EEG, he supplemented its
readings by resorting to a quite different approach. In what became
known as the Montreal procedure, he took advantage of the absence
of pain sensors in the brain by administering only a local anesthetic
that allowed patients to remain conscious throughout the operation. He
then enlisted their help in pinpointing the epileptogenic zones responsible for their seizures. Lightly touching exposed brain tissue with an

electrode, he asked patients to report their reactions. By such means he was able to elicit the telltale auras that accompany seizures and to zero in on precisely defined target areas, thereby minimizing the damage to other areas of the brain. Penfield's patients thus became collaborators in their own treatment.

Penfield described the sense of mystery that accompanied him in his earliest scientific explorations in tones of wonderment: What was going on, he asked, within "the living, thinking human brain on the operating table below me? . . . I stood before the functioning brain of man as before the great Sphinx and was dumb." Robbed of speech himself, he resorted to ingenious ways of making the brain communicate—or, as an article in the *Saturday Evening Post* put it in 1949, "vocalize"—first through EEG and then through his patients' utterances. That combination of methods, he wrote, revealed to the surgeon "many secrets about the living brain." On first hearing of Penfield's methods, his teacher Sherrington responded in jest, "It must be great fun to have the 'physiological preparation' speak to you."

In the course of probing his patients' brains, Penfield made a remarkable discovery: In response to electrical stimulation of the brain area known as the temporal lobe, some patients reported that they experienced vivid visual and aural sensations, including snatches of music and imagery of various kinds. They also experienced emotions, as well as, seemingly, memories that were often marked by an almost hallucinatory intensity and immediacy. Penfield was initially unsure of how to interpret the occurrences. Were they simply further aura-related phenomena? hallucinations? flashbacks of some kind? By the early 1950s, he had become convinced that they were actual memories—so-called hallucinatory reminiscences, phenomena first speculated about by the British neurologist John Hughlings Jackson in the late nineteenth century. They had, he theorized, lain buried or dormant in the brain but had now been reactivated by the stimulus of the electrode. He began referring to individuals who reported such instances of recall as his "memory patients" and devoted himself to gathering evidence to confirm his developing theory that all experience was preserved perfectly intact in memory and that it could, under the right conditions, be reproduced with near-absolute fidelity. Whether naturally occurring or artificially induced, seizures, claimed Penfield, replayed part of what the philosopher William James had called the "stream of consciousness," which left behind it "a permanent record that runs, no doubt, like a thread along a pathway of ganglionic and synaptic facilitations in the brain. This pathway is located partly or wholly in the temporal lobes."

Not only did Penfield's methods enable him to go beyond his predecessors in establishing the existence of such phenomena, but his account of the underlying process was illuminated by the metaphorical possibilities of midcentury technologies unknown to nineteenth-century scientists. When the neurosurgeon's electrode touched certain points on the cortex, he wrote, the reactivated memories resembled "a strip of film from a moving picture. It is not a still picture but a moving progression of images." The strips of film, he was convinced, were permanently available to the electrode. Over time, his views about the stability, permanence, and integrity of memory became increasingly bound up with the metaphor of the moving image. At the moment of its perception, the original experience was, like a film, "projected on the screen of man's awareness before it is replaced by subsequent experience." A record was then, so to speak, archived and available for replay, exactly like a cinematic flashback. Penfield's conception of filmlike records stored in the subconscious brain became basic to midcentury accounts of memory.*

Penfield's discoveries, as Oliver Sacks later put it, seemed to confirm Jackson's view that epileptic seizures were "portals" or "transports" to altered states of mind, including the "fossilized memory sequences" to which the English neurologist had given the name "reminiscences." One of the strangest qualities of those altered states was their doubleness. In certain cases, patients seemed simultaneously to relive the experience

---

* The emergence of the modern sciences of memory, Alison Winter has written, was closely tied to the appearance of new forms of mass media. Winter, *Memory: Fragments of a History* (Chicago: University of Chicago Press, 2013), 3.

and to report it to the surgeon; they were present both in the relived scene and in the operating room, thus experiencing the reminiscence as both actor and spectator. Though Penfield accounted for the phenomenon in terms of the stream of consciousness, other observers, as we shall see, found in it grounds to surmise the existence of more complex psychic mechanisms, including those of repression and the unconscious.

It was in Penfield's institute that the psychologist Brenda Milner, several years prior to her first encounter with Henry Molaison, first found employment (see Clinical Tale 1). On the recommendation of her doctoral supervisor, Donald Hebb, Penfield hired her to help assess the cognitive consequences of the operations he performed on his patients. She later wrote of being deeply impressed by the experience of sitting in the gallery of Penfield's operating theater as he stimulated a patient's exposed cortex, eliciting, in some cases, complex hallucinations, though she remained somewhat skeptical of the claims he made about them. Milner carried out extensive psychological tests on the patients before and after surgery, measuring intelligence, memory, and other functions, as well as conducting long interviews with them. Her findings remained limited, however, due to the nature of the constraints that Penfield imposed on himself. Committed to minimizing the losses suffered by his patients, Penfield removed only half of their limbic lobes. Any cognitive deficits they experienced as a result were thus limited to one hemisphere and could be compensated for by the intact structures of the other. It was William Scoville's lack of restraint—his operating bilaterally, thus eliminating the possibility of compensation—that made the case of Henry Molaison such a momentous one for the history of neuroscience.

The Information Paradigm

In many ways, as we have seen, the shift that took place in brain science in the 1950s crystallized in the studies Milner carried out on H.M. beginning in the mid-1950s. Prior to that time, as Eric Kandel has written, scientific thinking about memory and learning had been a "muddle"; researchers thought of the cerebral cortex as a "bowl of porridge" in which all regions were similar in function. For the most part the then-dominant fields of human science, behaviorism and psychoanalysis, simply ignored the brain, treating it as a black box. But the research carried out at Montreal's Neurological Institute and elsewhere began to give shape to a new conception of the brain as an organ whose pathways and functions could be explored and mapped. Penfield, who referred to the prevailing view of the brain at the time his career began as "an undiscovered country," played a major role in filling in many of the spaces in that uncharted terrain. With the help of his patients, whose exposed tissue lay open to him while he operated, he used electrical probes to identify key sites and pathways of motor and sensory function, thereby producing a map of what he called "functional localization." Up to a certain point this mapping remained within the parameters of a strictly physiological approach. But in the postwar era, as his interest in his "memory patients" grew, Penfield became increasingly focused on the question of "mind" and on bridging the gap between the "neuro" and "psy" disciplines. Parallel research projects further contributed to that endeavor. In the late 1940s, the psychologist Donald Hebb, his colleague at McGill, identified a key mechanism of learning dubbed the Hebbian synapse, an area of enriched neural connectivity (colloquially known by the phrase "Nerves that fire together wire together"). The recognition that processes of perception, learning, and memory had a neural basis—which Brenda Milner's studies did much to confirm—laid the foundations for the emergence of cognitive neuroscience.

By the 1960s and 1970s, behaviorism's grip over the human sciences was being loosened by an approach that pried open the black box and began treating its contents—consciousness—as an information-processing system. As with Penfield's cinematic analogies of memory, the emergence of that new informational model was accompanied by a reliance on new figurative language. Concepts of cerebral and mental functioning became infused with computational analogies. (Penfield's final work, *The Mystery of the Mind* [1975], is emblematic of this.) The language of circuits and programs became integral to the new approach and its effort to break the hold of the behaviorist prohibition on mind and brain. Over the course of the second half of the twentieth century, many aspects of what it means to be human—memory, perception, cognition, mental illness—were recast in the language of information.

That process was further reinforced by another landmark midcentury development that drew attention to the centrality of biological factors in mental processes such as perception, memory, and learning. Kandel traced that shift to the role that Francis Crick and James Watson's discovery of the helical structure of DNA in 1953 played in giving biology a new scientific prestige.* As Crick wrote at the time, the discovery "formalized information as a fundamental property of biological systems." In 1962, the year he and Watson received the Nobel Prize for their work, Crick invoked their discovery as heralding a paradigm shift in the life sciences, one that made it possible for the first time to begin to conceive of consciousness in biological terms. The question of the relationship between brain and mind, he predicted, would be one of the great scientific problems of the future. It is worth noting that the very notion of a scientific paradigm shift—a "revolution" in the way a given scientific community conceives of its primary object of investigation—was itself a product of that same moment. Thomas Kuhn's landmark book *The Structure of Scientific Revolutions* was published in 1962, and although its focus is on physics, his argument applies equally well to the life sciences. As Ian Hacking wrote in the introduction to the fiftieth anniversary edition of Kuhn's book in 2012, the Nobel Prize awarded to Watson and Crick in 1962 was a sign of things to come. If physics was then where the action was, fifty years later, with the Cold War over, the action had moved to the life sciences.

---

* In his memoir the Vienna-born Kandel related that, growing up, Sigmund Freud had been his intellectual hero and that his idolization of Freud had continued even after he had turned away from psychoanalysis to pursue a career in brain science. Though he devoted himself to studying the biology of memory (with the help of sea slugs), he nursed a lifelong belief in the continuities between Freud's ideas about memory and his own, suggesting that Milner's demonstration of the existence of implicit, or unconscious, memory "was the first time a scientist had uncovered the biological basis of a psychoanalytic hypothesis." Eric R. Kandel, *In Search of Memory: The Emergence of a New Science of Mind* (New York: W. W. Norton, 2006), 133; on Crick and Watson, xi.

From Neurophysiology to Psychoanalysis
and Back Again

If midcentury psychoanalysts, like most behaviorists, largely tended to ignore the workings of the brain, the quest for a biological basis for Freudian hypotheses remained a tantalizing possibility for some scientists. One such was the psychoanalyst Lawrence Kubie. Though his name is not widely known today, Kubie is an important figure in this story. After receiving his early training in neurophysiology, he became a leading figure in American psychoanalysis. He held a position at Yale and ran a private practice in Manhattan. He was also an active collaborator in the scientific transformations of the period, participating in several landmark conferences, including the Laurentian Symposium held in 1953 (see preface).* Kubie nursed a strong desire to forge connections between his field and the emerging sciences of the brain. That desire went, for the most part, unreciprocated. Yet Penfield's interest in forging an integrated approach to the sciences of mind and brain meant that he was prepared to enter into wide-ranging cross-disciplinary collaborations. Along with the psychologists Milner and Hebb, the EEG expert Herbert Jasper, and the Harvard neuropsychiatrist Stanley Cobb, Kubie served as one of the consultants on Penfield's ongoing Temporal Lobe Research Project.

In the early 1950s, Penfield opened his operating theater to what may well be one of the most unusual experiments in the history of science. Fascinated by Penfield's account of his patients' recollective utterances while on the operating table, Kubie saw an opportunity to probe further, perhaps, as his friend Kandel later wrote, to discover where in the "elaborate folds of the human brain" the psychic agencies identified by Freud

---

* He also participated in the Macy Conferences, from which cybernetics emerged. Kubie coupled scientifically innovative with culturally conservative views. He treated homosexuality as an illness to be cured, forcing his patient Tennessee Williams to break off several homosexual relationships. See Eli Zaretsky, *Secrets of the Soul: A Social and Cultural History of Psychoanalysis* (New York: Alfred A. Knopf, 2004), 299, 313, 322.

might be housed. Penfield and Kubie first crossed paths at a 1951 symposium at which Kubie commented enthusiastically on Penfield's presentation of the latest findings of his research on his memory patients. Shortly thereafter he traveled to Montreal to sit in on several of Penfield's operations and subject patients to psychoanalytic free-association tests as they underwent surgery. In each of the cases, as he subsequently related, the aura preceding an epileptic attack was artificially induced by electrical stimulation:

> An Iowan farm girl of thirty-five was on the operating table for the extirpation of an epileptogenic focus in the temporal lobe. For nearly twenty years she had suffered from psychomotor seizures with automatisms and various psychological symptoms. I was under the drapes with my recording equipment making a verbatim record of her words throughout the course of the operation and the electrical stimulation of the temporal cortex.
>
> Shortly after the exposure of the temporoparietal areas, a series of pictures were presented to the patient. These were simple line drawings of familiar objects, which she was asked to name. She did this quite readily until Dr. Penfield stimulated the area in the lobe which caused an arrest of speech. This occurred just as she had been shown the drawing of a human hand. At the moment of electrical stimulation all speech ceased as though frozen into immobility. *It was like watching a motion picture brought to a sudden halt.* Then as the current was turned off she began first to make mouthing movements and then slowly and hesitatingly said in sequence: "five, five . . . five horses . . . five horses . . . five pigs . . . five pigs . . . five fingers . . . hand." (author's emphasis)

In the paper in which he published the results of that experiment, Kubie drew far-reaching conclusions. Positioning his views squarely within the newly emergent informational paradigm, he began by citing his intellectual debts to cybernetics and its founders, Warren McCulloch, Norbert Wiener, and Walter Pitts (who, in conversation with Kubie, once compared the unconscious to the appendix, opining that both were useless and prone to disease).* Kubie's paper is permeated by the language of circuits and feedback, as well as allusions to the efficacy of extreme interventions such as lobotomy in cases where psychotherapy came up against its limits. The value of such shock therapies, he wrote, lay in the fact that by interrupting "negative loops" they could render the individual accessible to new psychological influences and his or her own inner controls. He stressed the need for multidisciplinary research along those lines.

From there he proceeded to a discussion of Penfield's discovery that the temporal lobe was the storehouse of experience and that perfectly preserved flashbacks or filmlike traces of past experiences could be activated through electrical stimulation of the area. The importance of that discovery to psychoanalysis was clear, he wrote, given that "almost everything we know about the neurotic process . . . is closely linked to the fate of hidden memories." Yet whereas Penfield confined himself to claims about the realm of conscious (if forgotten) memory, Kubie leapt to speculations about the storage of unconscious, repressed memories. He wondered whether the limbic lobe, sometimes referred to as the visceral brain, might not in fact be "the psychosomatic organ"—the physiological locus of one of Freud's psychic agencies. Moreover, he found reason to connect one particularly intriguing feature of the experiences reported by Penfield's memory patients—the phenomenon of double consciousness, of being both actor and spectator—to the psychoanalytic theory of dreams:

---

* In a footnote he recorded an exchange with McCulloch, who in a letter to him stated that the problem with Freud's model of the nervous system was that it remained tied to nineteenth-century concepts of energy, while current models stressed the nervous system's role in processing information. The nervous system's major function, as Kubie put it, was "the reception, organization, coding, filing and reassembling of informative messages from both the internal and external environment." Lawrence S. Kubie, "Some Implications for Psychoanalysis of Modern Concepts of the Organization of the Brain," *Psychoanalytic Quarterly* 22, no. 1 (1953), 24–25.

To quote Penfield: "The song goes through the mind, as he heard it on a certain occasion"; or the subject finds himself in a situation which progresses and evolves just as in the original situation; or it is "an active and familiar play, in which he himself is both the actor and the audience," the subject feeling "doubly conscious of the two simultaneous situations." I heard, for example, one patient on the operating table say as her temporal cortex was stimulated, "Now I hear you double," and later, "Now you are far, far away—all unreal." Note here how closely this resembles our dual role in a dream in which we are both the observer and the observed. . . . Also, note the suggestion of unreality feelings, depersonalization, and other dissociative phenomena. Frequently, the electrically evoked experience "was more vivid than an ordinary memory." Certainly this is the very stuff of dreams and symptoms, evoked in these instances on the operating table.

Ultimately, continued Kubie, Penfield's findings and the cinematic metaphors in which he couched them could be regarded as the equivalent of hypnotically induced regressions into the past. If Kubie was correct in his speculations, new vistas were opened up for coordinated research by "psychoanalysts who are sophisticated in neurophysiology and by neurophysiologists who are sophisticated in psychoanalysis." Pondering his encounters with Penfield's verbalizing patients who were apparently reliving long-buried experiences, Kubie envisioned a major transformation of the talking cure in which the psychoanalytic couch would yield to an entirely new approach. Electrical stimulation of the temporal cortex might, he surmised, evoke in a few moments exactly the kind of reexperiencing of the past that the analyst struggled for months or years to achieve.

Whatever modification of psychoanalytic method that implied, Kubie believed that such modification would be more than compensated for by the overall strengthening of the field's position: "It seems to this author that such considerations should ease the qualms of those neurophysiologists who still boggle over the term, 'unconscious,' as applied to psychological processes. Now that a neurosurgeon has demonstrated on the operating table the existence of this 'library of many volumes,' unconscious yet dynamically charged with the lifelike vividness of sensation and affect, the fears of our more psychophobic colleagues should at last be set at rest." Unfortunately for Kubie, such hopes proved premature.

In a letter he wrote to Penfield in early 1954, Kubie enthused that "very few people have been privileged to hear the spontaneous reports of patients responding to electrical stimulation producing—to paraphrase your terms—the technicolor strip of previous experience. Even fewer of those persons have been psychologists, and fewer still have had the experience, as have I, of witnessing comparable 'strips' brought up during psychotherapy sessions." Yet though he seemed to wholly adopt Penfield's cinematic metaphor, Kubie also sought to revise Penfield's model in keeping with his own psychoanalytic commitments. To illustrate that, he included a diagram of a rectangle whose corners marked four respective positions: (A) operating table, (B) couch, (C) operator (surgeon), (D) psychotherapist. It was, he suggested, Penfield's belief that the memory "strip" at A was authentic (i.e., a completely faithful "replaying"), while the "strip" at B was less apt to be so. Yet regardless of whether the stimulus was electrical or emotional, he argued in a direct challenge to Penfield's ideas, the resulting strip was likely to be based on a composite memory. As examples he cited both a composite childhood memory of his own and that of a patient who, in a sudden rush of emotion, had experienced a moment of vivid recall: "Then, as if watching a movie of herself, she comments: 'I am climbing the back stairs of our house . . . etc.'" On subsequent questioning the patient had conceded that she might have fantasized certain details of that "movie." Pressing a point he had made to Penfield at the Laurentian Symposium in August 1953, Kubie challenged Penfield's view that the content of the strip itself was arbitrary and of little inherent significance. That, he suggested, might be explained by the different forms of stimulation employed: the emotion-laden strip produced at B (couch) was likely to be more meaningful than that produced by the arbitrary nature of the electrode at A (operating table). The patient at A, he suggested, thus needed to undergo the process at B to "help shed further light on this question." Given the impossibility of asking psychoanalytic patients or doctors to expose their cortex to stimulation, he concluded that hypnosis might be the only method that might allow comparison of the memories produced at these two respective positions.

What did Penfield himself make of Kubie's formulations? There is no record of his response, but there is reason to believe that he remained skeptical. In letters to their mutual friend the Harvard neuropsychiatrist Stanley Cobb, Kubie struck a somewhat plaintive note. Asking Cobb for his impressions regarding Penfield's attitude toward psychoanalysis, he wrote that Penfield's earlier interest in collaboration seemed to have cooled. Cobb's response was guarded, seeming to place some of the blame for Penfield's attitude on Ewen Cameron, the chairman of the department of psychiatry at McGill University and a controversial figure to whom we will return later. The relationship of the Montreal Neurological Institute to psychiatry in Montreal, he confided, was poor because of Cameron, a man he described as "shy of both psychoanalysis and neurophysiology." More to the point was his surmise that Penfield was not convinced that control experiments on the unconscious were possible. As one scholar has put it, Penfield's interest in mind did not extend to the unconscious. Yet if, from Penfield's perspective, Kubie may have read unwarranted meanings into observed phenomena, so, too, scientists have subsequently argued, was Penfield guilty of overinterpreting his findings. Later research suggested that the memories Penfield thought he elicited were simply the auras or hallucinations associated with epileptic seizure. The trouble may already be seen to exist in his reliance on a naive model of the filmlike properties of memory.* Brenda Milner, for one, voiced reservations about that model that partly echoed Kubie's, writing that Penfield's "tape recorder" notion of memory "seemed highly implausible since I had been trained to think of memory as a reconstructive rather than a reproductive process." Despite those reservations, she remained a colleague of Penfield and thus, as she later wrote, was able to ensure the survival of neuropsychology at the MNI. The same was not true of psychoanalysis.

---

* Memory, as Oliver Sacks later put it, is not a static entity but a dynamic process. Sacks, *Hallucinations* (New York: Vintage, 2012), 155.

Psychoanalysis at Midcentury

Notwithstanding the somewhat embattled tone of Kubie's remarks, the position of psychoanalysis at midcentury was quite strong by comparison with its prewar situation. One of Kubie's close allies, Roy Grinker, had done much to help bring that change about by virtue of his success with treating psychological casualties during World War II. Grinker adopted a modified form of the talking cure to help psychoneurotic soldiers recover memories of the battlefield traumas that had precipitated their breakdown. Other military doctors similarly experimented with approaches ranging from hypnosis to group psychotherapy. As a result of its wartime contributions, the Freudian approach won considerable prestige in the United States and emerged as a major voice in postwar debates about the reorganization of the mental health profession.[*] That marked the beginning of a period stretching from the late 1940s to the mid-1960s in which psychoanalysis became a dominant presence in American psychiatry. Remarkable though it now seems, by the end of that period more than half of all psychiatric departments in the country were headed by figures with at least one foot squarely in the psychoanalytic camp.

Much of the new authority enjoyed by psychoanalysis was based on its wartime service. War had dramatically transformed attitudes toward psychoneurosis, which was initially treated dismissively by both the military and the medical establishment. Penfield's own attitudes are illustrative: prior to his postwar reorientation toward psychology and the question of "mind," he viewed most war neurotics as malingerers, an attitude shared by many of his peers. But as had already happened in

---

[*] The cover image of a 1948 issue of *Time* magazine depicted the psychoanalyst Karl Menninger, the newly elected president of the American Psychiatric Association (APA), alongside an image of the human brain, an image signaling both the new recognition accorded the field of mental health care in postwar American society and the success of psychoanalysis in establishing a major claim over that field. That claim, however, did not go uncontested, leading to pitched battles between the Freudians and their more somatically oriented professional rivals. See Andrew Scull, *Madness in Civilization: A Cultural History of Insanity* (Princeton, NJ: Princeton University Press, 2016), 339–40.

the previous world war (though its lessons had largely been forgotten) the epidemic scale and urgency of the problem of wartime psychiatric illness opened the door to new approaches. The American psychoanalytic community mobilized to fill the void created by the shortcomings of traditional asylum-based psychiatry, whose approach remained largely fixated on the problems of severely psychotic patients and shaped by a belief in the hereditary basis of most mental illnesses. War, writes the historian Ellen Herman, normalized psychological breakdown as well as the psychoanalytic deemphasis of predisposition. Psychoanalytically trained doctors such as Roy Grinker and John Spiegel, who coauthored the important text *Men Under Stress*, pioneered new methods of treatment that adapted Freudian methods to the battlefield. As the North African hospital where they were stationed was flooded by traumatized soldiers, they began experimenting with a method known as narcoanalysis, which involved administering sodium amytal or sodium pentothal to help the soldiers recover memories of the battlefield traumas that had precipitated their breakdown. Like hypnosis, another technique widely used during the war, narcoanalysis was believed to work by disabling the psychic mechanism that resulted in repression of the traumatic memory.

As the historian of science Alison Winter has written, the memory project that Grinker and Spiegel became associated with thus rested on assumptions similar to those of Wilder Penfield: namely, that memories were perfectly preserved records of experience, stored like the frames of a film—an analogy that they, like Penfield, frequently invoked. Though in many respects the two fields remained far apart, with regard to the key question of memory they converged around a belief that it was preserved in the form of intact, stable records and that the recall of these experiences could be compared to the experience of watching motion pictures.

On the basis of its success with employing an accelerated form of talk therapy, psychoanalysis emerged from the war with its professional status greatly enhanced. Freudian practitioners began extending the lessons of wartime to the problems of postwar civilian society, an endeavor that found affirmative popular expression in films such as Alfred Hitchcock's *Spellbound*. Soon its claims to diagnose and treat mental illness spilled from the private consulting room into the wider social and political arena.

America's psychoanalytic community diagnosed a wide array of alleged psychosexual developmental disorders in postwar society, from juvenile delinquency to homosexuality to the so-called undemocratic personality. Freud's teachings were thus enlisted to help diagnose the political pathologies of Nazism and communism. Closer to home, they were also formative for analyses of the "paranoid style" in US politics and the new methods employed by the advertising industry's "depth men" (see chapter 5). At least in part, that new authority was based on an implicit ethical claim rooted in its stance against shock treatment as an unwarranted intervention into the mind and the resulting loss of memory.

The release in 1952 of the first edition of the *Diagnostic and Statistical Manual of Mental Disorders* (DSM) bore many traces of the ascendance of psychoanalysis. But if that decade marked the high tide of psychoanalysis in America it also marked the advent of a very different approach to mental illness whose emergence would first shake and then eventually dislodge psychoanalysis from its central position in American society. The year 1953 saw the introduction to the market of the antipsychotic chlorpromazine, an event described by one scholar as "a seminal moment in human history." Sold to the public under the name Thorazine, it was prescribed more than 2 million times in the year following its appearance. Its immediate effect was to challenge somatic treatments such as ECT and lobotomy.* For patients suffering from schizophrenia, it offered a form of treatment that acted on the brain without doing so invasively, and in the wake of its appearance the use of such methods went into precipitous decline. Its longer-term effects were every bit as significant. Despite its serious side effects, it contributed to a gradual emptying out of asylums, whose population had swelled to enormous proportions in the preceding decades. The process of deinstitutionalization radically changed the landscape of mental health care. That transformation was accompanied by the appearance of a broader class of psychopharmaceuticals that

---

* Following the introduction of Thorazine, the practice of lobotomy disappeared virtually overnight in California state hospitals. See Joel Braslow, *Mental Ills and Bodily Cures: Psychiatric Treatment in the First Half of the Twentieth Century* (Berkeley: University of California Press, 1997).

offered new treatment for depression and other syndromes. Inevitably, those developments also challenged the psychodynamic paradigm, laying the foundations for the later emergence of what some scholars have called a new understanding of the self in biochemical terms. That form of "biomedical selfhood" became fully manifest with the publication in 1980 of the third edition of the *DSM*, or *DSM III*, which dispensed with most Freudian classifications. By the end of the twentieth century, psychiatry was, according to many accounts, predominantly a brain science. To its critics, this had the effect of leaving the field virtually "mindless"; proponents of that development, on the other hand, responded by arguing that their opponents had embraced a perspective that, by radically deemphasizing biology, wound up "brainless."

Sciences and Fictions of Mind, Brain, and Memory

Developments in the midcentury sciences of mind and brain inspired many fictional accounts.* Penfield's findings were widely reported in the press, which documented his explorations of what one piece called "the fantastic geography of the brain cortex" and his finding that electrical stimulation created situations in which the "patient seems to be thinking with two minds." Such reports made their way into popular culture, indirectly inspiring the brain-in-a-vat scenarios that became a staple of science fiction in the 1950s. In many such tales, the technological apparatus and metaphors of midcentury science took on a strange, mutant life of their own. Some of those accounts also raised philosophical and ethical questions about the scientific practices featured in their pages.

Connections such as those also feature in the story of Ewen Cameron. A major figure in North American psychiatry, Cameron served as the president of the APA and head of the Allan Memorial Institute of

---

* In Stanisław Lem's 1961 novel *Solaris*, for instance, scientists visit a planet enveloped in a substance that seems to be endowed with consciousness. Operation Brain-wave, as their mission is called, involves experiments on this "ocean-brain," whose attributes seem to include the symptoms of epilepsy. The planet also conducts its own experiments on the scientists, who begin experiencing strange hallucinations.

psychiatry at McGill University. Yet within the rich scientific milieu that existed in Montreal, he remained something of an outlier, a figure whose antagonistic relations to neurophysiology and psychoanalysis were reciprocated by the withering judgments of Penfield, Hebb, and others. (Penfield's hope of integrating psychiatry into his interdisciplinary project foundered on his poor relations with Cameron.) Cameron seems to have responded to those professional slights by intensifying his efforts to leave a mark on the field, albeit in ways that ultimately mired his legacy in deep scandal. He became notorious for his research into experimental treatments (some of it, as we'll later see, CIA funded), though much of that research emerged quite naturally out of a utopian strain of midcentury psychiatry that sought to find a "grand cure" for all mental illness. In Cameron's case such ambitions manifested themselves in an entirely unique kind of memory project: the pursuit of methods of systematic erasure of patients' memories. Cameron experimented widely with most of the methods available to midcentury psychiatry: sodium amytal, LSD, prolonged (sometimes for weeks) sleep, sensory deprivation, and ECT. He used the last method highly aggressively, arguing that the memory loss that was one of its chief side effects—which led many other practitioners to use it as sparingly as possible—was therapeutically desirable in that it aided in the erasure of destructive behavior patterns held in the memory "storage systems." But that was only part of the cure, for Cameron sought to replace the erased memories (which, in his own adaptation of information theory, he called "tapes" or "programs") with new ones.

To do so, he enlisted the aid of an idea with origins in the world of science fiction, a genre that Cameron read voraciously. Introduced in a work of early-twentieth-century pulp literature as the "hypno-bioscope," the method was rechristened by Aldous Huxley in *Brave New World* (originally published in 1932 and updated in 1958 as *Brave New World Revisited* with an expanded discussion of the topic) as "hypnopedia" or sleep learning. Cameron commissioned an EEG specialist to design a device called the Cerebrophone and adapted it as part of his method of "psychic driving," a process of erasing patients' memories and then "implanting" new messages that were delivered to the patient in the form of taped loops that played endlessly while the patient slept. Whereas Penfield believed that memories were stable and well preserved, Cameron proceeded from the opposite premise that they could be radically altered, even permanently erased, a premise that would become emblematic of Cold War research into the malleability of memory and selfhood.

The case of Cameron demonstrates that for some practitioners the recovery of memory, understood as a straightforward veridical reproduction of past events, was not always seen as the most efficient path of return to mental health. The British psychiatrist William Sargant was another physician who used narcoanalysis to treat World War II servicemen, albeit in a form that modified the methods of Grinker and Spiegel. Sargant believed that patients could become "conditioned" to neurotic behavior and that in such cases the only way to restore them to health was to temporarily switch off their brains through prolonged sleep. He reported that in some instances he encouraged drugged patients not to relive their earlier experiences but rather to imagine new ones, similar though not identical to the original traumatic events. An implanted memory, he felt, could trigger a more intense emotional discharge than a real memory could. Cameron and Sargant were representative of a group of midcentury "psychiatric adventurers" who experimented with destroying memories and implanting new ones in order to rebuild the self. At the far end of this story we thus encounter a body of memory practices quite different from those associated with narcoanalysis in its original form: practices in which fictive memories or implants become central to conceptions of recovery and selfhood. In them the metaphorical film associated with the process of recall was, so to speak, scripted and directed by the physician.

Sargant and Cameron were far from alone in their willingness to explore radical solutions to their patients' illnesses. As contemporary brain scientists sought to lay claim to the terrain of personhood and establish their methods as the new measures of health, memory, and identity, some of the most serious contests in midcentury American society were fought out on that terrain. The emergent paradigm of the explorable brain, it was felt by many, had deeply ambivalent implications. By the early 1960s, in a rapidly changing social climate, a backlash was taking shape against the no-holds-barred experimentation of the 1950s.

As Rebecca Lemov has written, "the civil rights movement was bringing about a . . . rise in consciousness of the rights of human subjects
in medical and psychological experiments." Lobotomy, ECT, "psychic
driving"—all were swept up in a wave of moral, political, philosophical
revulsion and critique.

Those methods also found their way into a rich vein of philosophical
speculation, as we see if we return in closing to Wilder Penfield's operating theater and the Montreal procedure that he pioneered there. The
historian of science Cathy Gere has connected that procedure to a strand
of intellectual history that extends from the seventeenth-century French
philosopher René Descartes to the philosopher Hilary Putnam's thought
experiment of the "brain in a vat." Updating a scenario first imagined by
Descartes in the 1600s, in which he contended with the radical doubt
that his knowledge of the external world was an illusion crafted by an
"evil demon," Putnam asked his reader to imagine the science fiction figure of a mad scientist controlling all inputs into a disembodied brain
that is suspended in a vat (a scenario that represents yet another of the
mutant offspring of midcentury brain science, appearing in 1950s sci-fi
and horror at the same time as the first press accounts of Penfield's work;
for more, see chapter 5). "By varying the program," wrote Putnam, "the
evil scientist can cause the victim to 'experience' (or hallucinate) any situation or environment the evil scientist wishes. He can also obliterate the
memory of the brain operation, so that the victim will seem to himself to
always have been in this environment." The question Putnam then posed
is: How do we know that what we know is true?*

Gere traces that scenario to a century-long tradition of experimental
medicine, much of it involving patients with epilepsy. Despite the limits
Penfield himself placed on the Montreal procedure, both the methods he
pioneered and the metaphors he drew upon point to Putnam's thought
experiment. Though they began with epileptic subjects, subsequent

---

* In a slightly earlier version, noted Gere, two philosophers in 1968 speculated about
the possibility of a so-called braino machine, a device worn on a subject's head that would
produce hallucinations.

experiments with electrical stimulation of the brain eventually broadened to encompass the possibility of behavioral modification on a much wider scale, a possibility that, not surprisingly, generated deep unease. Putnam's thought experiment, suggested Gere, "emerged in the context of a widespread cultural panic about the possibility of electrical stimulation of the brain." In that context Descartes's "evil genius" was replaced by a "slightly unbenign" incarnation of the figure of Wilder Penfield. In a strange twist, Penfield's operations and the claims about memory they led him to make turned out to undercut their own premises: in the brain-in-a-vat scenario, the contents of thought and memory are elicited, and perhaps entirely produced, by an external influence or agent. Such possibilities became part of the wider popular reception of the midcentury revolution in brain science. Within a context marked by a growing cultural anxiety about the fragility of memory and identity, Penfield's claims for the integrity and stability of memory were called into question, not simply by his methods of eliciting memories through electrical probes but also by his reliance on the metaphors of the mass media that became central to midcentury accounts of "brainwashing."[*]

---

[*]   At about the same time Putnam was proposing his thought experiment, Wilder Penfield wrote that (a) the brain was a computer and (b) it must be programmed or operated by an "external agent," which, in a dualist conception similar to that of Descartes, he identified as the mind. Descartes's error, he noted, lay in identifying consciousness with the pineal gland. Penfield, *Mystery of the Mind: A Critical Study of Consciousness and the Human Brain* (Princeton, NJ: Princeton University Press, 1975), 60, 109.

OPERATORS AND THINGS

Memory, madness, and midcentury technical analogues for psychic processes came together powerfully in Barbara O'Brien's memoir *Operators and Things: The Inner Life of a Schizophrenic* (1958), an account of a descent into mental illness that became a touchstone for the midcentury antipsychiatry movement. O'Brien (a pseudonym) described a world divided into two categories of being: Operators, those who had agency, and those, like herself, who did not and were therefore Things. She traced her discovery of that state of affairs to a personal breakdown triggered by office politics that produced states of dread and paranoia that she compared to those described by Orwell in *1984*. She experienced the male-dominated office where she worked as an extended assault on her person and her memory. Like the bureaucrats in Orwell's novel who use the "memory hole" to rewrite the past according to the needs of the present, she became a collaborator in a similar process that ultimately resulted in her own complete break with reality. "By the time you have twisted the facts to agree with the picture you wish to see," she wrote, "your subconscious mind has helpfully plodded through the past and distorted a lifetime of facts to make them agree with your present self-deception program."

That discovery precipitated a crisis in which O'Brien came to believe that she had been selected for participation in an experiment carried out by so-called Operators, figures equipped with special "brain cells" that allowed them to peer into and control the minds of Things such as herself. Fleeing her job and hometown, she embarked on a mad journey, crisscrossing the country by Greyhound bus in an effort to escape rival groups of Operators who competed for control over her psyche. The resulting highly episodic narrative combined elements of noir, science fiction, and pure whimsy. Characters have names such as "the Spider" and "the Western Boys" in a manner that seems to cross Lewis Carroll with William S. Burroughs. Repeated assaults are launched on the contents of her mind through a process variously described as "draining," "chiseling," or "scalloping" that is designed to destroy memories and induce amnesia; behind that process lurks the specter of shock treatment or psychosurgery.

In addition to hypnosis, telepathy, and other such methods, the Operators exercise influence over Things by means of an array of technological contrivances. One such device is the stroboscope, which the Operators use "to probe, feed in thoughts, take out information, or keep a watching eye on the person." (According to O'Brien, the devices could operate for distances of up to a mile.) Here we see how the methods employed by contemporary brain scientists such as Grey Walter entered mass culture, becoming objects of popular knowledge, fascination, and unease. At one point in the narrative, the narrator's stay in a psych ward becomes the setting of an experiment that involves administering so-called blackout movies, which are designed to erase all memory of the Operators from the narrator's mind. The situation becomes ambiguous, however, because although technological apparatuses enter the narrative as manifestations of malign agencies acting on a besieged self, elsewhere they become part of her recovering mind's self-representation.

As the narrator gradually frees herself from the grip of the Operators, her fantastical tale resolves itself into the more or less familiar terms of the topography of the conscious and unconscious. The consequences of this split are conceived of in cinematic terms: "Your mind is 'split,' and a subconscious portion of it, no longer under your conscious control, is staging a private show for it." The paranoid's world, she wrote, "is a wide-screen movie." In the words of author Mike Jay, "The world of operators and things is analogous to a movie directed and screened by one part of her unconscious mind to keep her occupied while another part quietly repaired the trauma." Motion pictures, as O'Brien's account makes clear, were not simply a ready, even obvious analog for mental functions such as memory (per Penfield); they had become integral to the way people at midcentury conceived of their inner life. Penfield's notion of "double consciousness," in which the patient simultaneously experiences and observes the "hallucinative recollection," is extended and deepened in O'Brien's account of her experience as director, actor, and spectator of a film.

In its demonstration of the way in which figurative language may take on a life of its own, O'Brien's narrative seems to bear out the critique of Penfield's analogies voiced by Karl Lashley, a leading contemporary scientist of memory. At the meeting of the American Neurological Association at which Lawrence Kubie had commented enthusiastically on Penfield's presentation of the findings of his work with his memory patients, Lashley had offered a more skeptical response, calling into question Penfield's fondness for analogies between media and neural processes and arguing that such analogies were problematic not least because they conveyed an impression of knowledge that was at best premature, if not wholly erroneous. Lashley remained unconvinced that the phenomena described by Penfield were in fact memories. He also spoke of what he called "a curious parallel in the histories of neurological theories and of paranoid delusional systems." Since the days of Franz Mesmer, he reminded his audience, paranoiacs had been tormented by each successive generation of new technology, right up to present-day telephone and computer systems; their delusional systems simply built on the scientific predilection for facile parallels between nervous and technological systems.

In yet another analogy that emerges toward the end of O'Brien's narrative, the power of the unconscious, for both good and ill, takes on a cybernetic dimension. Alluding to hypnopedia and the wondrous feats of memory that are attainable through it (a topic discussed by Aldous Huxley in his contemporaneous book, *Brave New World Revisited*), O'Brien wrote:

> The unconscious has a similarity to the huge electric brains which have apparently been patterned on the more obvious abilities of the unconscious mind. Into the electronic brains are fed tapes of detailed information and in the brain is installed machinery for processing the tapes. . . . I had always thought of the unconscious as a whirling pool of repressed emotions, better repressed. Instead, it appeared to be a sort of private Univac, an incredible piece of thinking mechanism, the possession of every conscious mind on earth.

Virtually nothing is known about O'Brien apart from the fact that at the end of the six-month-long psychotic episode documented in her book, she made a full recovery. Tracked down twenty years after the book's publication, she chose to maintain her anonymity. The book itself, however, was quickly recognized as a classic of its kind. It was widely reviewed and praised by, among others, R. D. Laing, the British antipsychiatrist whose ideas about the positive meaning of the hallucinations associated with schizophrenia were influenced by O'Brien's account. Indeed, *Operators and Things* seems in some respects to have been written as a direct contribution to the literature of midcentury antipsychiatry.

That movement emerged at the same moment in the 1950s that the emptying out of US asylums began. That process, which came to be known as deinstitutionalization, was a major development in the history of mental health care, driven by many factors, among them the invention of a new class of antipsychotics that enabled patients to be treated on an outpatient basis. It was also driven by shifts in the funding of mental health care away from the states, which sought to unburden themselves of the drain on their budgets posed by the ever-growing asylum populations, to the federal government, which embraced the—as it turned out, chimerical—ideal of community-based mental health treatment. At the same time, it reflected the emergence of a powerful critique of psychiatry, particularly in its asylum-based and aggressively somatic form, which was perceived to have turned the brain and psyche into a site of invasive experimentation. That critique swept across much of Western society in the postwar era, galvanized by the writings of charismatic figures such as Laing in England, Michel Foucault in France, and in the United States the sociologist Erving Goffman, known for his unsparing view of the asylum as a "total institution," and Thomas Szasz, a renegade physician in whose eyes the very notion of mental illness was a "myth." Their views became very influential, though their influence was often exaggerated by critics who blamed them for the debacle of deinstitutionalization, as the failure of governments to fully fund the community clinics meant to replace the asylums, coupled with the tendency of many psychotic patients to go off their medications because of their unpleasant side effects, resulted in an ever-growing population of untreated mentally ill homeless individuals. Perhaps somewhat more substantively, opponents of antipsychiatry argued that in its critique of the "mindlessness" of midcentury psychiatry, it had embraced a perspective that, by almost entirely de-emphasizing biology, had wound up "brainless."

It is not difficult to see why O'Brien's account would be embraced by antipsychiatrists such as Gregory Bateson and Laing. It serves not least as an example of how illness narratives may offer insight into the operations of contemporary society, a society that seems to enforce a conspiracy of silence against the patient's voice.* This insight must be

---

* Antipsychiatry was a complex phenomenon that spanned the political spectrum, from a left-wing critique of capitalist society to a right-wing version in which it became essentially synonymous with anticommunism; to those in the latter camp, psychiatric treatment shared many features with Red "mind control." One of the most successful variants of the latter form was Scientology, which, under the leadership of the sci-fi author L. Ron Hubbard, advertised itself as a humane alternative to the invasive practices of modern psychiatry. Scientology attracted a large following that included a cross section of the Western cultural intelligentsia, including figures such as Aldous Huxley and William S. Burroughs. (For more, see Clinical Tale 6.)

won back from doctors, who seek to deny it; as portrayed in O'Brien's narrative they are the ultimate enforcers of the conspiracy. Her story is notable not simply for its protest against the ethical lapses of a profession in which the (frequently female) patient all too often becomes research subject but for its assertion of the patient's right to voice her own account of madness. In its self-presentation as an extended road trip across midcentury America, it also celebrates, however ambiguously, the new possibilities of existence beyond the confines of the asylum.

Yet O'Brien's narrative is by no means a straightforward brief for antipsychiatry. Though the narrator is haunted by the specter of the Ewen Cameron—like physician and comments bitingly on the many head doctors, including psychoanalysts, she encounters along her journey, the pulp terms of her account ultimately distance her from the image of the "Mad Scientist" (along with its inevitable accompaniment, the brain in a vat). When that apparition surfaces in one of her hallucinations, she simply laughs it off. Though critical of the way that contemporary medicine treated illnesses of the mind as illnesses of the brain, she does not go to the other extreme of favoring, like the antipsychiatrists, a view of the self as "brainless"; she expresses deep fascination with theories about the chemical and hormonal causes of schizophrenia. Nor, in the end, does she provide much comfort to the partisans of psychoanalysis. O'Brien's is a psychological crisis, acted out in the terms of pulp melodrama: "How very odd," she muses, "that my unconscious mind should call itself an Operator and call my conscious mind a Thing." Her hallucinations are caused neither by Penfield's electrode nor by Walter's stroboscope but arise from internal conditions—not a chemical imbalance but an "emotional upheaval." Yet this is not just a self-portrait but a portrait of a highly technologized society in which ideas about selfhood have become permeated by the mass media and by technical contrivances for influencing it.

PART II

LIEUTENANT DULLES

The young officer lay in a coma for more than three weeks. During an attack by North Korean forces on the evening of November 15, 1952, he had suffered a traumatic head injury when a projectile had entered the right lower, or parieto-occipital, part of his skull and penetrated into the rear left area of his brain. Upon recovering consciousness, he found himself in a hospital in Yokosuka, Japan, to which he had been transported by ship following the injury. Several operations had cleared out metal and bone fragments and relieved the pressure that had been building in his cranium. His speech returned haltingly at first, but as his convalescence progressed over the month of December, he began to give voice to fragmented memories and anxieties.

The patient seemed haunted by fears that he had acted dishonorably or gone insane. His statements were both rambling and obsessive. "I beg your pardon sir," he said to the hospital's commanding officer, "I don't want to appear to doubt your word, it's only that I don't want to get the idea that I did nothing dishonorable, only to find that it's not true." "You see," he continued, "my mind is incapable of any recollection and although I do not wish to doubt your word, it is of utmost importance to me not to get the wrong impression." Anxiously, he persisted in questioning the doctor: "You say I didn't become insane? You see I am very much afraid I became insane and then was sent back—You say I was wounded at the front?"

Many of these statements were recorded by his mother after she flew to Japan to join her convalescing son. As she noted in her journal, the confirmation that he had been wounded rather than lost his mind came as an enormous relief to the young officer. Yet he continued to express confusion, oscillating between moments of relative calm and others of abject terror. In a journal entry recorded by his mother on December 20, 1952, he seemed to hallucinate the presence of his absent father: "Oh father, I've been so horribly used. Will you hold me by the hand so that you won't lose me? . . . I desperately need you—I'm so terrified." Again he called out, "Father can you take me into your house? Father, can you make sure I don't get kidnapped again. . . . Where can we go to be safe? I've been so terribly long a prisoner. I'm desperately afraid, so much has happened in the past." "Father," he pleaded once more, "you must remember I'm psychologically afraid."

The belief that something terrible had happened to him—being wounded, "used," "kidnapped," taken "prisoner"—was coupled in the young man's mind with the sense that he had done something terribly wrong. Again and again he returned to his fear of having acted dishonorably: "My mind is very evil now." He had an obscure intuition of having been caught up in dark events in which he had unwittingly played a part. A fragmentary entry jotted down by his mother recorded him asking point-blank, "Of what am I accused?" Prey to nameless fears, he felt an awful sense that the very ground of his being was slipping away: "Mother could you tie me with a cord so that I won't disappear. I'm awfully much afraid that if I go to sleep I'll disappear."

Allen Macy Dulles, the young officer in question, was the twenty-two-year-old son of Martha "Clover" Todd Dulles and Allen Welsh Dulles. Fresh out of Princeton, where he had graduated at the top of his class, he had enlisted in Korea, refusing the easier posting available to him as a son of one of the most powerful men in the world. Even while young Allen returned to the United States to continue his convalescence, Dulles senior was appointed the director of the CIA, a position in which he oversaw US intelligence operations at the dawn of a new era of intense ideological conflict—a "battle for men's minds," as it was often called, or, as Dulles himself called it in a speech he gave to a gathering of Princeton students in May 1953, "Brain Warfare."

Eyewitness accounts later confirmed that the younger Dulles had been acting with exceptional bravery when he suffered his injury, putting himself into the line of fire in an effort to save several of his men who were pinned down by enemy snipers. Why, then, was he plagued by fears of having acted dishonorably? Why did he refer to being kidnapped, held prisoner, and "used"? Those fears may simply have been the delusions of a man suffering from severe head trauma. Yet they have a striking relationship to the great drama being acted out in Korea that became central to the campaign of "brain warfare" overseen by his father. The alleged brainwashing of American POWs by their Korean and Chinese captors would become the object of a cultural panic in American society and a central element in a narrative about what was at stake in the Cold War. In his tormented ruminations the younger Dulles seemed to allude to the specter of collaboration that haunted accounts of what had been done to US servicemen in captivity.

Allen Macy Dulles never fully recovered from the head injury he suffered in Korea. Following his return to the United States in February 1953, he began a medical odyssey that would last for decades, taking him to hospitals across the United States, Canada, and Switzerland. Around the case of the son of America's top spy gathered a formidable array of talent in the fields of neurosurgery, neurology, and psychiatry, among them Wilder Penfield, Harold Wolff, and the Swiss psychiatrist Ludwig Binswanger. Some of those figures—such as Wolff, a neurologist at Manhattan's New York Hospital—were recruited into Dulles senior's new program of clandestine research in clinical and behavioral science even while they were consulting on his son's treatment.* Young Dulles underwent repeated tests, examinations, and treatments, including insulin shock and antiepileptic medication, as well as a weeklong study by Herbert Jasper, the EEG specialist at the Montreal Neurological Institute. As Penfield wrote to Dulles senior, fragments in his son's left temporal lobe probably accounted for the hallucinations that plagued him, and the EEG might shed light on that. Following Penfield's subsequent decision to take the patient off medication (Dilantin), he began experiencing seizures and was plagued anew by hallucinations and paranoid obsessions. Some of them concerned family ancestry and genetics; others, relating to the atom bomb, the Jews, and the tormenting fear of collaboration, are a window into the dark climate of that time.

The confusion surrounding young Dulles's symptoms meant that he was often treated as a "mental case." That led to psychiatric hospitalization as well as a course of Jungian treatment, continuing a family connection that dated back to World War II, when his father, the head of the Office of Strategic Services (OSS), the precursor of the CIA, in the European spy capital of Bern, Switzerland, had asked the Swiss psychiatrist C. G. Jung for a psychological profile of Hitler. But he obtained no real relief until the appearance of a new generation of antipsychotic medications in the 1980s helped him regain partial functionality and autonomy. His inability to form stable memories, however, kept him living in a permanent twilight state. As if already intuiting the amnesia that would haunt him for the rest of his life, one of the diary entries made by Allen's mother in late 1952 included her son's poignant wish: "Make a note—not to forget me."

---

* At least one biographer has conjectured that Allen Welsh Dulles's interest in "brain warfare" was driven in part by his son's traumatic brain injury. Peter Grose, *Gentleman Spy: The Life of Allen Dulles* (New York: Houghton Mifflin Harcourt, 1994), 396. See also Stephen Kinzer, *Poisoner in Chief: Sidney Gottlieb and the CIA Search for Mind Control* (New York: Holt, 2019), 134–35.

The story of Allen Macy Dulles contains numerous echoes of the case histories of other patients whose tales have entered the annals of modern medicine. As we saw in the case of Henry Molaison, memory and its loss play a central role in many such stories, as it did in the subsequent history of brain science, a field then beginning to undergo revolutionary changes, many of them galvanized by research on patients who, through war or other injury, had suffered catastrophic impairment. These tales are part of the origin story of modern neuroscience and a window into a profoundly transformative historical moment.

Significant features of the historical constellation that took shape in the 1950s are encapsulated in Allen Macy Dulles's story. In particular, his personal tragedy sits at the intersection of two stories central to the narrative of this book: that of the midcentury transformation of brain science and that of the rise of what his father called "brain warfare." Often told separately, these two stories overlap in many ways. The entanglements between them played an underappreciated role in the late-twentieth-century emergence of the field of neuroscience. The seeds of those developments were planted in the period of scientific ferment that occurred at the dawn of the Cold War, a period in which the brain was studied, experimented on, and imagined in a host of new ways—many, as we will see, intimately related to the novel forms of warfare central to that conflict.

One of the technological *a prioris* of this story, as we saw in chapter 1 and as Allen Macy Dulles's stay at the MNI reminds us, was EEG, which helped usher in a new era in brain science. Such devices soon also became part of the arsenal of Cold War behavioral science; a memo written by his father in the early 1950s identified EEG, along with electric shock and the polygraph, as part of a repertoire of techniques whose utility in the field of interrogation was to be investigated by CIA personnel. This, then, is not simply a story about the rise of a new era of technoscience. It is equally a story about changes in the nature of politics and warfare. The Cold War put the psychological and behavioral dimensions of geopolitical conflict front and center. The specter of minds both friendly and alien, but also of the confusion

surrounding this most basic of distinctions, loomed large throughout that period. New paradigms of information, communication, and
control that emerged from World War II spawned, as Dulles senior put
it, unconventional forms of warfare.* "Man himself," rather than arms
or supplies, became the critical factor in war. War, politics, and society were reconfigured in light of the new significance accorded this
conception of man, his capacities and vulnerabilities. The sciences
of brain and behavior assumed a central role in the process, a role
actively promoted by Dulles senior. His appointment as head of the
CIA three weeks following his son's return to the United States began
the decade that came to be known as "the Dulles era" at that agency.
Dulles's carefully cultivated image of himself as a gentleman spy, observed Tim Weiner, masked a considerably darker reality. In addition
to the numerous coups that he authorized (in Iran, Guatemala, and
elsewhere), Dulles oversaw a program of clandestine experimentation
in interrogation techniques, culminating in the early-1960s classified
manual *KUBARK Counterintelligence Interrogation*, a document that came
to public light in the early 2000s in the wake of the torture scandal at
Guantánamo Bay. *KUBARK* may be seen as a direct outcome of Dulles's
reframing of the Cold War in terms of its central icon, the Iron Curtain:
that, as he put it in a speech, was both a "geographical barrier" and
a "state of mind," a symbol of "the most complete and all-inclusive
thought control program in history," which the Kremlin sought to
impose on the peoples of the Eurasian landmass living between the
Elbe River in the west and the Chutotka Peninsula in Russia's far east.
To combat that menace, Dulles authorized the clandestine program
known as MKUltra, which sanctioned and funded a vast enterprise of
research into the field of "brain warfare" (see chapter 4). The secrecy
surrounding that program, which was deemed necessary to protect
the American public from gaining knowledge of what its own government was doing, illustrates a fundamental contradiction that became

---

* For media theorist Marshall McLuhan, the Cold War was, quite simply, an "information war."

institutionalized during the Dulles era: that between democracy and secrecy.*

The brain's entry into midcentury political history was also reflected in works such as Hannah Arendt's influential book *The Origins of Totalitarianism* (1951), whose analysis borrowed at key points from contemporaneous developments in behavioral science. In stressing the threat to public knowledge and memory posed by totalitarian states and the alleged Pavlovian methods they deployed, Arendt echoed the warnings of novelists such as George Orwell and Aldous Huxley. In the preface that Huxley wrote in 1946 for a new edition of *Brave New World* (originally published in 1932), he predicted a postwar world marked by what he called a "Manhattan Project" of behavioral science, a "revolution of body and mind" with ominously antidemocratic implications. Orwell explored similar themes in *1984*. Such texts became integral to the Western public's understanding of what was at stake in the Cold War in terms of a battle for men's minds.

But what did it mean to invoke a battle for men's minds? If the phrase often remained an abstraction, it was made literal in the specter of brainwashing that haunted both Dulles junior and Dulles senior. Amid the panic that seized American society in response to reports of the phenomenon, accounts of strange new forms of human vulnerability became integral to the Cold War imagination. The story of US servicemen seemingly subjected to mysterious methods of interrogation, manipulation, and indoctrination became a clinical tale writ large, a fate served up as an object lesson to the American public. But it was not their fate alone; in many midcentury accounts the American public itself, beset by new forces seeking to mold thought, loyalty, and behavior,

---

* The distorting effects of this secrecy doctrine on public accountability are illustrated by the congressional testimony provided by Richard Helms, Dulles's eventual successor as CIA director, in the 1970s. Asked to reflect back on his predecessor's dark legacy, Helms, wrote Thomas Powers, wound up sounding like an amnesia victim: "It is said men begin life with a *tabula rasa*; Helms ended it that way." Powers, *The Man Who Kept the Secrets: Richard Helms and the CIA* (New York: Knopf, 1979), 379.

was treated as a kind of clinical case. Scientists were accordingly recruited to devise new techniques for probing the collective mind—a kind of EEG, it might be said, suitable for use on a mass scale.

During the 1950s, accounts of the kind of traumatic battlefield injury and memory loss suffered by Allen Macy Dulles were often accompanied by reports of the uncanny new forms of psychic injury associated with the campaign of clandestine warfare waged by his father. Indeed, there are hints, in the fragmented fears voiced by young Dulles—particularly in the way he seemed to assign himself a role in a captivity narrative or larger drama—of that other kind of warfare. As we shall see, the line between clinical syndrome and political pathology was often blurred during that era: in the years to come, anxieties about the integrity of brain and mind would frequently take on an explicitly political cast.

CHAPTER 3

An Experiment with Mankind

*In Red China all of the accumulated techniques of years of
experimentation on the Soviet people are now being used. The
results, as [Edward] Hunter observes, are producing a sort of
"Alice Through the Looking Glass" national mental picture.*

—DR. LEON FREEDOM

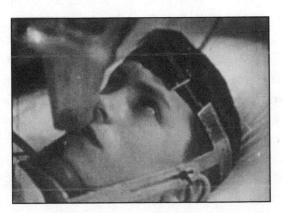

Vsevolod Pudovkin, The Nervous System
(Mechanics of the Brain)*, 1925.*

REPRINTED WITH PERMISSION OF GOSKINO.

In his 1956 book *Brainwashing: The Story of Men Who Defied It,* the
journalist and former intelligence operative Edward Hunter described
an obscure Soviet film he had uncovered while researching background

material for the book. Titled *The Nervous System* (or *Mechanics of the Brain*) and made by the Soviet director Vsevolod Pudovkin in the mid-1920s, the film was a work of scientific popularization that introduced the physiologist Ivan Pavlov's research on conditioned reflexes to its audience.* As the film demonstrated, Pavlov had discovered that automatic responses could be created in the dogs he experimented on through the repeated association of a primary stimulus such as food with a secondary stimulus such as a light or bell. The film seemed to suggest that by the mid-1920s, Pavlov—perhaps, as Hunter implied, at the behest of Vladimir Lenin himself—had extended his experiments in reflex conditioning to human subjects, a development that, if true, had deeply disturbing implications.

What exactly did Hunter see in that obscure production? Why did he endow it with the elements of a horror film and at the same time identify it as the key to the Reds' campaign of "brain warfare"? The story of its introduction into midcentury accounts of Communist mind control, which Hunter had first described five years earlier in his book *Brain-Washing in Red China: The Calculated Destruction of Men's Minds*, is one that tells us as much about the anxieties and uncertainties of American public life of the time as it does about the nature of the Cold War. In Hunter's telling, his discovery of *The Nervous System* seemed to pierce the veil of darkness surrounding one of the deepest mysteries of the Cold War: it helped illuminate the methods that, according to him, had been used on American POWs in the recently concluded conflict in Korea. In a way that no mere written or verbal account could, it provided graphic evidence of the reality of brainwashing, a central, albeit highly ambiguous, Cold War construct.

Cold war, as Hunter and others came to define it in the early 1950s, was essentially synonymous with psychological warfare. Hunter himself had plied his trade in that field since World War II and would continue to do so up into the 1960s. As a systematized body of knowledge and

---

* Pudovkin became one of the USSR's leading filmmakers, directing films such as *Storm over Asia* (1928).

practice, its origins were quite recent, though over the course of World
War II it had assumed a major strategic role that grew only more significant in the ideological confrontation that developed between the
superpowers from the late 1940s on. As defined by Hunter's longtime
acquaintance Paul Linebarger in his influential text *Psychological Warfare*
(1948), psywar was "continuous war"; warfare, according to Linebarger,
was now to be understood as simply a particularly violent form of a
broader and ongoing campaign of "persuasion" that characterized all of
modern politics (for more on Linebarger, see Clinical Tale 6). Authorities
in the field argued that its emergence reflected a historic transformation
in the nature of warfare in the modern age. In the twentieth century, as
the objective of war shifted from control of territory to control of population, questions of morale, belief, and loyalty moved to the forefront
of strategic considerations. In 1941, as part of a survey of psychological
warfare commissioned by the nation's wartime Committee for National
Morale, the Hungarian émigré Ladislas Farago wrote, "Total war inevitably made man himself (his attitudes and sentiments), rather than arms
and supplies, the focal point for determining ultimate victory or defeat."
Such a view was echoed by many leading American statesmen.

The Cold War provided numerous variations on the theme of war
as a struggle for men's minds—a form of warfare in which the field of
battle shifted from an external to an internal terrain and was waged as
much with words and images as with guns. Yet defined in those terms,
cold war represented an abstraction, a conflict without conventional bor-
ders or forces. It remained a nebulous concept, seemingly inadequate to
the task of mobilizing national resolve and resources. In the early 1950s,
however, it assumed a new and more vivid form. Hunter's writings, beginning with his 1951 account of brainwashing, played a central role in
that development by giving concrete expression to what was at stake in the
new form of conflict. By 1953, the freshly appointed CIA director, Allan
Welsh Dulles, could confidently assert the existence of what he called,
in a speech given that spring (and later published in *U.S. News & World
Report*), "Brain Warfare—Russia's Secret Weapon." Recent revelations
about developments in the Soviet bloc, he told his audience, confirmed
that the Soviets had transformed the still young art of psychological

warfare into something without precedent in human history. In the area under Soviet control, "a vast experiment is under way to change men's minds": "Its aim is to condition the mind so that it no longer reacts on a free-will or rational basis but responds to impulses planted from outside." Obliquely referencing Pavlov's methods, he went on to describe the Soviets' use of techniques for creating new "brain processes and new thoughts": "Parrotlike the individuals so conditioned can merely repeat thoughts which have been implanted in their minds by suggestion from outside. In effect the brain under these circumstances becomes a phonograph playing a disc put on its spindle by an outside genius over which it has no control." Here we encounter another version of the brain-in-a-vat scenario previously discussed in connection with the work of Wilder Penfield: a disembodied brain, severed from its environment, whose perceptions are entirely controlled by the inputs fed into it by a mad scientist. This is no coincidence, for the possibilities explored by the famed neurosurgeon and others like him had become part of the way in which the brain was conceived and imagined by the public at midcentury. That image, as we shall see, haunted the discourse about brainwashing.

Though here Dulles cited the example of the phonograph to illustrate how the new style of warfare had reconditioned brain patterns, later in the speech he invoked another form of the mass media to make his point. Earlier Soviet experiments in the field had culminated, he went on, in the treatment of American POWs in Korea:

> The Communists have recently been showing a film portraying young American aviators who publicly make spurious "confessions" of participation in the use of germ warfare against North Korea. We have a copy of this film and I saw a showing the other day. Here American boys . . . stand up before the members of an international investigatory group of Communists from Western Europe and the satellites and make open confessions, fake from beginning to end, giving the details of the alleged dropping of bombs with bacterio- logical ingredients on North Korean targets. They describe their indoctrination in bacteriological warfare, give all the details of their missions, their flight schedules, where they claim to have dropped

the germ bombs, and other details. As far as one can judge from the film, these pseudo confessions are voluntary.

Those staged events and the fictitious accounts they were meant to publicize bore "the usual hallmarks of Soviet-imposed fabrications."

The six-month period, Dulles continued, that generally elapsed between the date of capture and the date of confession allowed for the drafting of an elaborate scenario containing the details of the confession and the time needed for successful conditioning of the "patient." In that period of time, a cinematic fiction, a kind of mass hallucination, was substituted for reality.

As we shall see in the next chapter, Dulles's speech coincided with his authorization of the classified program MKUltra, the Americans' version of the Soviet bloc's "mind control" program, though in fact, the United States had for years been developing its own capabilities in this field and a 1951 JCS document had invoked the Manhattan Project as a model for this program. Here we shall focus on the Americans' development of their own scenario and on the role that a particular representation of the brain played in it. Part I of this book explored midcentury developments in the sciences of brain and mind; in part II the focus shifts to the entanglements of politics and science: the ways in which, to varying degrees and varying effects, those sciences became woven into a vital strategic narrative about the emerging superpower conflict.

Though Dulles alluded to earlier demonstrations of the Soviets' capabilities in the field of "brain warfare" (e.g., the Soviet show trials of the 1930s), the immediate occasion for his speech, and the crisis around which American public consciousness of brainwashing crystallized, was the case of US servicemen taken prisoner during the Korean War. The first signs of trouble appeared early in the war, when in 1950 a captive US officer made a broadcast on the enemy's behalf. Far worse were the developments of 1952. In a major public relations coup for the Communists, thirty-eight US Air Force POWs signed confessions of their involvement in an alleged germ warfare campaign, an event captured on film and transmitted to a global audience. The results, as Dulles's speech made plain, were a disaster for the US forces. They confirmed earlier accounts that hinted at the eerie developments unfolding on the Korean Peninsula.

From 1950 to 1953, the Korean War—one of many civil wars of the immediate post–World War II era—served as the setting of the first direct confrontation between the Cold War foes. In many ways that conflict became the Cold War's defining event. It set into motion developments that defined not just geopolitical realities but also domestic realities for the rest of the century and beyond. Not the least of them was the fact that it laid the foundations for the national security state that emerged in the United States during the 1950s. That development institutionalized a state of emergency that put the country on a more or less permanent war footing, turning it, in the words of the political scientist Harold Lasswell, into a "garrison state." More specifically, Korea became the first test of NSC 68, the expansive new foreign policy doctrine formulated in 1950. The policy of containment, of seeking to check communism in strategically important regions while avoiding direct engagement, was set aside for a more aggressive stance meant to confront directly the new nemesis around the globe. The sources of this shift were rooted in domestic politics and anxieties as much as in actual threat levels; the latter remained hard to gauge and the subject of intense speculation. The lack of a solid empirical basis for much of this speculation—concerning, for instance, the existence of a so-called missile gap—lent the effort to forge a Cold War consensus a phantasmal quality. US officials often addressed this problem by framing the conflict in terms that drew on works of the imagination. Though social scientific works such as Hannah Arendt's *The Origins of Totalitarianism* (1951) played an important role in producing knowledge of the enemy, figures such as the statesman George Kennan were frank in acknowledging their debt to fictional portrayals of what was at stake: "When I try to picture totalitarianism to myself . . . what comes into mind most prominently is neither the Soviet picture nor the Nazi picture as I have known them in the flesh, but rather the fictional and symbolic images created by such people as Orwell or Kafka or Koestler or the early Soviet satirists." The problem of knowledge that such statements allude to crystallized in the phenomenon of "brainwashing."

The Korean War was conducted amid a pervasive atmosphere of

unreality, one that continues to haunt it until today. Sensitive to war-weary public sentiment, President Harry Truman refused to call it a war, preferring the Orwellian euphemism "police action." To this day it remains the least known of modern wars for Americans. So complete is this sense of not-knowing that the default name for it, according to David Halberstam, is "the forgotten war." Perhaps it is an effect of this that it has received far less cultural attention than the nation's other wars. According to the historian Bruce Cumings, this is less the result of neglect than of deliberate forgetting, a kind of actively produced amnesia or repression complete with its own fictive or implanted memories. Indeed, much of the war, including many of its most unsavory aspects, has been erased from Americans' collective memory. This includes the brutalities of the South Koreans, the United States' allies, who carried out systematic massacres of North Korean villagers and prisoners, as well as America's own atrocities, above all a massive air campaign that resulted in the deaths of more than 3 million North Koreans (estimates are that a third of the peninsula's total population died). The 1951 firebombing campaign against the capital, Pyongyang, dropped enormous quantities of napalm on the city. Cumings has shown how some of those deeds—among them a massacre of several thousand North Korean villagers by South Korean troops while US troops stood by and watched—were converted into Communist atrocities through a filmic campaign that completely inverted the true course of events.

One of the central functions of the emerging national security state, as such incidents illustrate, was the management of public perceptions. Indeed, the assumption of such a function by this new entity was one of the preconditions of its birth. This was an unprecedented development in US history, one that aroused considerable misgivings at the time (President Dwight D. Eisenhower, one of its major architects, was still voicing them at the end of the 1950s in his warnings about the "military-industrial complex"). In their efforts to sell NSC 68 to Congress and the public, US officials adopted the tactics used by advertising firms and spoke of the need for a psychological "scare campaign." The problem, in David Halberstam's words, was that the "American century" was about to begin but no one wanted to pay for it. It was amid those circumstances that the Korean War, as Secretary of State Dean Acheson would put it, "came along and saved us," ensuring the final approval of NSC 68 and a fourfold increase in US military spending (from 4.7 to 17.8 percent of GDP) at a time when the American public favored a substantial cutback in the resources committed to the military. According to Acheson, "June 25 [1950] removed many things from the realm of theory." The invasion of South Korea seemed to provide direct confirmation of the necessity for NSC 68. What had been veiled in speculation now became fact.

But what exactly was fact in the context of a war that could not be called a war? For the historian Bruce Cumings, the story of the conflict in Korea cannot be told without reflecting on the fragile nature of public memory and knowledge in modern America. The question it posed, he wrote, is: How do we know that what we think we know is true? That enigmatic war/nonwar confirmed that "We who live in Western liberal society have our subconscious automatically (if imperfectly) produced from birth." Amid the climate of institutionalized secrecy and paranoia that hung over the birth of the national security state, a climate greatly intensified by the malignant effects of McCarthyism, much about that conflict remained shrouded in mystery and half-truth. Henceforth American security, and the awesome apparatus created to defend it, would depend as much on the management of public perceptions, emotions, and memory as it did on weaponry. The mass media naturally assumed a vital role in that ongoing task. Their role is central to the plot of the iconic film to emerge from the war, *The Manchurian Candidate*, a film that thematizes the "hegemony of amnesia" that, according to Cumings, surrounded both the war itself and the national security state it helped spawn (see chapter 5).

From certain perspectives, US firepower, however awesome, seemed almost to count for nothing in the face of evidence of the Communists' advantage in the field of "brain warfare." If the Cold War was to be fought as a war of managed perceptions, the great "brain operation" that, according to some, was under way behind the Iron Curtain showed that the Eastern Bloc was far ahead. It was probably for that reason that the brainwashed US POWs became the key dramatis personae of the conflict. Despite the enormous loss of life and the destruction wreaked across the Korean

Peninsula during the war, brainwashing emerged as the central plot point in a distinctly US-centered narrative. It became the vital lesson of what was at stake in that strange new faraway conflict—a lesson from which American commentators were eager to draw far-reaching conclusions for a public less than receptive to the news.

Of none was that truer than Edward Hunter. For Hunter the first radio broadcasts by US POWs in 1950 had torn open the veil over what was happening in the Soviet bloc. Later developments such as the germ warfare confessions in 1952 only reinforced his belief that something deeply ominous was happening in the POW camps. In his writings the camps were nothing less than laboratories for carrying out the ultimate in scientific experiments—an experiment designed to test the capabilities of the totalitarian state to mold people according to an entirely new pattern. The experiment, as he ultimately conceived of it, was designed in accordance with a blueprint whose origins could be traced back to the work of Pavlov. This theme was echoed by many observers at the dawn of the Cold War. No less a figure than Hannah Arendt lent her imprimatur to it when she defined the concentration camp as the central institution of the totalitarian state and wrote that it had served as the setting of the "ghastly experiment of eliminating, under scientifically controlled conditions, spontaneity itself as an expression of human behavior." Continuing later, she added, "Nothing then remains but ghastly marionettes with human faces, which all behave like the dog in Pavlov's experiments."* As we shall see, Pavlov would eventually become central to Hunter's portrait of the enemy and its master plan for world domination.

"Ahoy! The Brain!"

What is often missing from historical accounts of Americans' encounter with brainwashing is, strangely enough, the brain. Most accounts of the

---

* Here Arendt seems to have leaned heavily on the writings of the Sovietologist Robert Tucker. According to the political scientist Darius Rejali, it was the British psychiatrist William Sargant who first identified Pavlov as the "evil genius" behind Soviet interrogation.

period tended to focus on mind control and to posit it as an extension of psychological warfare. Any connection to the developments in midcentury brain science that we explored in previous chapters was either denied or simply ignored by the sober social scientists who came to dominate the discussion. That connection was, however, central to Edward Hunter's account. Though he himself did not always draw a firm distinction between brain and mind, the primary emphasis in his account remained the brain—an emphasis that only intensified over time, even or especially in the face of mounting skepticism. Hunter used the term *brainwashing* literally. That insistent literal-mindedness is manifest in the way he repeatedly described developments in the modern brain sciences to help establish a plausible case for a construct whose reality remained, from the outset, highly speculative.

Surprisingly little is known about Edward Hunter. His story has not yet found its teller, perhaps as a result of the elements of confabulation that are so basic to its outlines. Born in New York City, he worked for several decades as a journalist in the Far East and Europe before serving during World War II in the OSS, the precursor of the CIA. In the war's aftermath he resumed his career as a journalist while also continuing to maintain ties with the US intelligence community before eventually going on to become the editor of the leading psywar journal *Tactics*. Throughout that period he also served as an important figure within a network of staunchly anti-Communist figures, including Ayn Rand and Paul Linebarger, who adopted a totally uncompromising stance against what they saw as the vacillating tendencies within US government and society.

Hunter's first writings on the topic of brainwashing appeared in a series of newspaper articles in the year 1950. Those, as he never grew tired of reiterating, gave him a strong claim to paternity of the term, his translation of the Chinese phrase for "cleansing the mind."* Following his introduction of the term into public consciousness, the term took on

---

*   See Corr. 1930–1953, Edward Hunter Papers, for repeated exchanges with *Time* magazine over the question.

a kind of talismanic meaning for him—a meaning that he expanded on at greater length in his book *Brain-Washing in Red China* (1951), which became an essential text of the Cold War era. Subsequently the term *brainwashing* went on to acquire a life of its own, with meanings its originator could hardly have anticipated. Though often labeled a myth, it was a concrete myth, resting, as Darius Rejali has put it, on a "plausible but fictive chain of conspiratorial associations," one that was firmly rooted in the imaginative possibilities of its time and one that generated powerful reality effects. It became part of what Marshall McLuhan called in *The Mechanical Bride* (1951) the "folklore of industrial man."

Hunter's account of what he called the "brain operation" taking place in China was prefaced by a section titled "Ahoy! The Brain!" in which he sketched the scientific context of that development. Based in Hong Kong at the time of the Chinese Revolution in 1949, he had had contacts with dissidents and others escaping China in the wake of that event who provided the germinal seed of his book. It opened with a brief account of his interview with one such dissident, Chi, who reported having been subjected to an exhausting campaign of political indoctrination based on endless harangues on party doctrine and dialectical materialism. Among its Chinese subjects the process was variously known as thought reform, self-criticism, or the new colloquialism, brainwashing. In the course of his interview with Chi, Hunter related to his readers, he had suddenly been struck by a "weird" train of thought. His interlocutor's account of the process he had undergone, he related, had triggered in him a chain of associations with a series of seemingly unrelated events. One of them was his recent visit to a friend back in the United States, who was undergoing hypnotic treatment while recovering from a nervous breakdown at a sanitarium. Another was a speech he had heard in which a US diplomat formerly stationed in Manchuria had spoken of being arrested, interrogated, and pressured into signing a confession to imaginary crimes. The third of the associations involved an evening he had spent in New York City before leaving for the Far East. Intrigued by his friend's treatment, Hunter had attended a hypnotic stage performance, during which the performer had implanted a posthypnotic suggestion in the mind of one of the audience members.

What was the link among those seemingly unrelated occurrences? Puzzling over these associations, Hunter searched for the inner connection among them. The hypnotic cure, it dawned on him, was in key respects strikingly akin to "mind reform," a practice powerful enough—as the stage hypnotist's demonstration suggested—that it might conceivably induce subjects to "confess" to nonexistent crimes. Hypnosis subsequently became a central trope within Hunter's analysis of what was going on in China. The cover of the paperback version of his book described the book's contents in the following lurid terms: "The first revelation of the terrifying methods that have put an entire nation under hypnotic control." But hypnosis was merely one element of the mystery of brainwashing. In the pages that followed the opening vignette, Hunter went on to describe a series of related developments in the art and science of probing the insides of people's heads: psychosurgery, psychiatry, cybernetics, and allied fields. Though the bulk of the book was devoted to interviews with victims of the new methods being practiced in China, he took pains at the outset to sketch a brief albeit deeply unsettling portrait of the new horizons that those sciences had opened up. Older certainties and pieties about the mind, he argued, could no longer be taken for granted amid present developments. Indeed, at certain moments in the opening chapter he seemed to echo antipsychiatric themes, although Western practices quickly disappeared from view in the account that unfolded, which emphasized the perversion of science in the Soviet bloc. In that sense his work was thoroughly embedded within a new scientific appreciation of the centrality of the brain to contemporary conceptions of humankind.

The unsettling nature of the knowledge to which he introduced his readers formed a recurring motif of Hunter's writings on the topic. This lent his prose a distinctly stagy quality. Repeatedly, in this as well as subsequent writings and public statements, he alluded to the "weird feelings" the subject aroused in him, to an eerie sense of *déjà vu* that his encounters with victims of the process engendered. Listening to his Chinese interlocutor, Hunter reported that he had been struck by a "peculiar feeling" that "certain passages in his description of brain-washing recalled some previous experience of my own." As the impression that he had heard it before persisted, he told the reader, "I tried to probe beneath forgotten brain layers in my own head to search for what it was that made his words, and particularly the weird unnatural feeling they gave me, so familiar." As though he himself were gradually emerging from an amnesiac episode or hypnotic trance, he drew us into his own "brain processes": "The feelings that had come over me in that most modernized institution while talking to the psychiatrist were the same as those I felt as I listened to Chi's story: the same disquieting sense of probing into dangerous fields." "One recollection led to another," and he recalled the demonstration staged by the hypnotist, which in turn prompted the question of whether such techniques could not be used for political purposes. Something like that, in fact, he opined, seemed to have occurred in related cases, such as that of the Hungarian cardinal József Mindszenty, the staunchly anti-Communist prelate whose sensational trial in 1949, at which he appeared to have been drugged, had first conjured up for the West the specter of Red mind control.

No doubt this way of unfolding his tale served a strategic purpose: in so doing he anticipated the skeptical reactions of his reader to the fantastical elements of the story he related. Yet the repetition is striking: at times it seems as if Hunter was in the grip of an almost mechanical compulsion; his desire to author the definitive account is unsettled by the sense that he himself is not the primary or sole author.* Ultimately, he told his reader, he realized what the basis for the connection he had stumbled across was. As he continued listening to Chi, he "became convinced that these remembered incidents and bits of incidents, seemingly so far apart, fitted with what he was telling me to form the rough outline of a pattern. They all had something to do with controlling the brain." He then went on to elaborate:

> Our age of gadgets and electronics had discovered the brain, and we were learning how to manipulate it. This was something

---

* It is also worth noting that Hunter's narrative places himself at the scene of important precursors: hence his claim to have been in Berlin in 1933 for the Reichstag fire trial, at which the defendant, Marinus van der Lubbe, was alleged to have been drugged or hypnotized (see his 1958 Senate testimony).

drastically new, like the splitting of the atom, that had come upon this earth in the middle of the twentieth century. . . . We had known vaguely about the geography and resources of the brain before, as we had known that there was much more to matter than met the eye. But the brain, like matter, had been a divine creation that could not be tampered with without paying a dreadful price. . . . The discovery of ways in which the brain operates has led to the discovery of how to control its movements, a tremendous new field of science.

This account of the mapping of the brain's inner geography was accompanied by a shift to a more exotic foreign geography: "My journey to the East had coincided with fascinating stories in the press in America about the construction by our mathematicians and mechanics of a mechanical brain that by use of electronics could compute in a moment what Einstein in his prime, with all his genius, might have needed many years to compute. Surgeons now are capable of extremely delicate brain operations that only a few years ago were literally impossible." Without mentioning their names, he invoked accounts of the exploits of Wiener, Penfield, and Walter that had appeared in the popular press in the years 1949 to 1950, immediately preceding the publication of his book.

What followed was a bold conceptual leap: "Man has learned not only some of the theoretical processes that go on in a man's head but also how to direct his thoughts, and to do this in a 'democratic group discussion,' in a 'self-criticism' meeting, on the operating table, or in the hypnotist's chamber." Modern psychology now encompassed everything that influenced thought and attitude, from publicity to psychological warfare, the modern communications media, public opinion surveys, and aptitude testing. Such techniques, he wrote in the conclusion to the opening section, now formed an inescapable part of contemporary economic and public life, not least in the field of politics: "The politicians of the world have been quick to seize upon these discoveries in the realm of the brain in order to advance their own objectives." They had thereby opened up vast possibilities in the field of psychological warfare, i.e., "warfare with unorthodox weapons."

Conditioned Man, or *Homo pavlovius*

Ivan Pavlov made no appearance in Hunter's first book on brainwashing. When he followed up that account five years later, however, in his book *Brainwashing: The Story of Men Who Defied It* (1956), Pavlov became the central protagonist of his new analysis of what was going on behind the so-called Bamboo Curtain. What had changed in the interim?

Though his first book, *Brain-Washing in Red China*, had been a bestseller, specialist opinion about Hunter's account remained decidedly mixed. Many experts expressed doubt that any esoteric or scientific practices had been involved in the confessions being extracted from US POWs, arguing instead that a combination of deprivation and coercion was sufficient to produce the desired results. Some, though, concurred with at least certain aspects of Hunter's account. Among them, as we have seen, was the political philosopher Hannah Arendt. For Arendt, the essence of totalitarianism was the effort to change human nature, to seize hold of the self at a deep level so as to remake it according to a new pattern, the model for which had been provided by Pavlov's dog experiments.

Claims such as those were echoed in studies commissioned by government agencies and think tanks. Throughout the early 1950s, Sovietologists, psychiatrists, and other experts debated the meaning of the disquieting developments associated with the Korean War and the significance of Pavlov in those developments. One key figure was the Soviet expert Robert Tucker, who argued that the Kremlin, faced with the challenges of making good on its claim to create a "new Soviet man," had in the postwar period found a new "formula of man" in the work of Pavlov. According to the new direction in Soviet science, "*The formula for man was the conditioned reflex.*" Its objective was to make the great physiologist's research on the workings of the nervous system directly applicable to the immense political tasks facing communism in its struggle against the West. In the 1930s, he wrote, Soviet psychologists had critiqued American behaviorism as a reactionary and literally mindless creed that reflected capitalism's need for a working class without any consciousness whatsoever; now they had switched to critiquing "idealist" forms of psychology such as Freudianism, to which they opposed Pavlov's "cortico-visceral physiology," a materialist doctrine that placed the nervous system and its role in mediating relations between body and mind at the center of the human sciences. That approach encompassed Pavlov's second signal system, the linguistic dimensions of human conditioning (according to which speech could cause deep changes in the whole organism). As part of the "Stalin-Pavlov" line now adopted by the Soviet state, the reflexes of Soviet citizens were to be "conditioned or reconditioned" by means of the state-controlled media.*

Whatever credence such accounts may or may not deserve—and they were treated skeptically by many in the US intelligence community—they achieved much currency in a political climate in which developments on the ground had created an urgent demand for analyses capable of explaining the seemingly inexplicable. In particular, the startling events of 1952–1953 generated renewed explanatory pressure on the early accounts provided by Hunter and others. In 1952, twenty-three US pilots being held in North Korean POW camps publicly confessed to war crimes as a result of their participation in an alleged campaign of bacteriological warfare. The confessions, as Dulles had observed in his "Brain Warfare" speech, were elaborately scripted and orchestrated spectacles that culminated in a filmic production that played to great effect before a worldwide audience. Worse was to follow. In the course of Operation Big Switch, the massive repatriation operation that followed the conclusion of hostilities in September 1953, twenty-one US servicemen refused to return home. Ideological conversion had now become a shockingly real fact, one that posed urgent questions for US government and society alike. The possibility that young American soldiers might find elements of the Communist worldview compelling simply could not be admitted. In

---

* The quasi-cybernetic overtones of that formulation became part of the way US intelligence officials conceived of the Soviet enemy. Richard Helms, an eventual successor to Dulles as CIA director, referred to the Soviet regime's efforts to mold its citizens into a "cybernetic state." The paragon of the "New Soviet Man," as scholars have noted, was the cosmonaut, whose training included courses in hypnosis designed to instill the necessary control over autonomic functions such as breathing in order to manage the physiological stress of liftoff.

response to the crisis, a small army of specialists was mobilized. Teams of experts in the clinical and behavioral sciences descended upon the more than four thousand POWs who ultimately returned home. According to one scholar, the US POWs became one of the most extensively studied groups in US history.

Disagreement about how to interpret the phenomenon of ideological conversion remained intense in US intelligence circles. Later studies suggested that the troops sent to Korea had made up "one of the least motivated armies America had ever put into the field." Problems of morale, however, were insufficient to explain the nature and scale of the crisis. Among those who rose to meet the challenge was Edward Hunter. He had reacted to the skepticism that had greeted his earlier account by blaming it on Red disinformation campaigns, complacent disbelief, or Communist sympathies. Yet even those who were favorably disposed to his views acknowledged that brainwashing remained a phenomenon beyond the capacity of average Americans to grasp. One such figure was Samuel Dean, an American engineer based in China who was arrested after the revolution in 1949 and subjected to harsh treatment before eventually being released. Hunter devoted a section of his new book to Dean's experiences. In July 1953, however, amid the relief that accompanied the negotiations to end the hostilities in Korea, Dean had written to Hunter to praise his first book but also to caution that its subject remained "as hard for ordinary Americans to understand as though it were brought back by a return [sic] Space Traveler from Mars." He advised that Hunter "bombard" the US public with a campaign waged through "newspapers, magazines, TV, radio and the like." Only by fully exploiting the mass media, he believed, would this story from Mars succeed in fully penetrating the "consciousness of American people." Indeed, official reports of the period tended to dismiss brainwashing as science fiction.

It was probably out of some such desire to heighten his book's impact that Hunter entertained the possibility of making its narrative, which often reads like a film treatment, into a "dramatic film." No such project ever materialized, yet his appreciation of film's possibilities in the field of psychological warfare is abundantly clear in the emphasis he

came to place on the early Soviet production *The Nervous System*. His
new book placed that film at the center of its further elaboration of
the perils confronting the complacent West. Deeply frustrated with
that complacency, he redoubled his search for new forms of evidence
of both the reality and the provenance of brainwashing and found it in
that obscure production.

Like his first book, *Brainwashing: The Story of Men Who Defied It*
opened with a discussion of the scientific basis of mind control. Now,
however, his focus was squarely upon Pavlov. He again began by noting
the "eerie sensation" evoked by the term *brainwashing*, the sense of coming up against something weird, transgressive, almost supernatural—a
sensation that created understandable horror and disbelief in the minds
of many. Precisely that quality worked against acceptance of its existence.
The Reds, he wrote pointedly, would prefer us to believe that there is no
such thing as brainwashing, and most Americans—whether acting in
good faith or bad—played along. Yet the phenomenon was all too real,
with demonstrable roots in a strain of modern science directly traceable
to the research of Pavlov. Hunter echoed Tucker in arguing that Pavlov's
work had experienced a revival in the Soviet Union at exactly the time of
the Korean War. Could this be simply a coincidence?

He then related how, as he had been researching background material for his new book, an acquaintance had revealed to him the existence
of a film set in Pavlov's laboratory that he had seen as a young medical student. It included scenes of the famous dog-conditioning experiments, in which new reflexes had been created by means of flashing
lights. But the film, his friend informed him, also included a scene with
a young man being subjected to similar experiments. That scene had left
such a vivid impression on him that he still experienced a "peculiar feeling" whenever he thought about it. Excitedly, Hunter realized that the
film might hold the key to the message he wanted to convey in his new
book. After much searching, he succeeded in locating a copy (probably
at Columbia University's Center for Study of Mass Communication).
He arranged to screen it in the presence of a small group of select
friends who included Ayn Rand, one of the presiding spirits of midcentury anticommunism, and his close associate Dr. Leon Freedom, a

neuropsychiatrist who became a key figure in Hunter's efforts to piece together the Pavlovian pattern that lay behind brainwashing. In the presence of that company, Hunter watched the film and reported on its contents in his book. One scene, captioned "the pathway of the arc of the conditioned reflex," included a pen-and-ink cross section of the dog's head that showed little gears ("a significant touch") connecting eyes and mouth with the brain triggered by flashing light. Soon the light itself was sufficient to trigger a salivary response even in the absence of food. The dog had "learned." (The accompanying caption read "Reflex caused by flashing light.") Conditioned reflex action, he explained, was what the Communists relied on to create the so-called new Soviet man. Disappointingly, however, the version of the film that he saw lacked the telltale scene of the young boy, no doubt, he surmised, because the Soviets had altered the original version. As part of its Orwellian strategy of rewriting history, the Kremlin had erased from public memory the scene that had revealed the film's true meaning, while releasing an edited version that celebrated a sanitized account of the great scientist's findings. Confronted with that erasure, Hunter's task was clear: to restore the missing scene to public memory and thereby expose the film's true significance within the Soviets' grand strategy.

After months of further searching, he was able to track down a complete version of the film and arrange a new screening. Seeing the missing scene for the first time produced in him a vivid reaction. He reported experiencing "a twinge of horror"—before quickly adding, "The twinge was involuntary, what Pavlov would have called an unconditioned reflex." The scene showed a young man strapped into a chair, being subjected to the identical conditioning experiments as the dog (see image on page 79). Eventually, as with the dog, flashing lights alone were enough to activate the boy's salivary glands. The implications that Hunter drew were momentous: the film revealed that literally any human activity, from salivation to romantic embrace to murder, could now be "clinically predetermined in politico-medical laboratories." Nowadays, in keeping with its emphasis on what Pavlov identified as the "second-signal system," the Kremlin used as stimuli not lights but words ("imperialism," "running dog," etc.), staged trials, and public confessions that were often captured on film and then screened for mass audiences. *The Nervous System* was thus not merely an illustration of Pavlovian doctrine but an enactment of the role played by the media in the process of political reflex conditioning. The human subject in the film was a kind of surrogate for the audience. Film, the production revealed, had been directly enlisted in the grand project of creating a "new Soviet man" as part of a process of "cinefication" of the world according to Pavlovian principles; hence Moscow's production of feature films pop-ularizing Pavlov's work, which made the scientist "a sort of master magi-cian with occultlike powers over men's minds, the Merlin of dialectical materialism."

Yet Hunter was not done. As though drafting his own film script, he further embellished his account by conjuring up for his readers a key scene from Pavlov's life. An old man by the time of the 1917 revolution, Pavlov was famously hostile to the Bolsheviks. Yet he was treated with great respect by them and lavished with resources at a time of tremen-dous scarcity. One day in 1923, an extraordinary summons was issued to Pavlov by Lenin, who asked him to come to what Hunter described as "an interview that was to be decisive in history." At that meeting Pavlov was commissioned by the Soviet leader to write a summary of his life's work on dogs and its application to man. For three months he labored on the man-uscript, then delivered it to Lenin; after reading it, the great revolutionary sent a message expressing his heartfelt gratitude to the great scientist. It laid the foundation for the scientifically engineered transformation of human nature that Karl Marx had originally believed would come about spontaneously through revolution: "Pavlov's manuscript, which became the working basis for the whole communist expansion-control system, has never left the Kremlin." With that account of the origins of the se-cret manuscript written by Pavlov, Hunter delivered the masterstroke in his piece of paranoid storytelling, implanting an entirely fictitious scene within the historical record. The manuscript became, in his telling, the Bolsheviks' master text, charting their plans for world revolution and eventual domination. It marked the true beginning of "brain warfare"; with respect to the human factor in modern war, he wrote ominously, the event stood on a par with the discovery of nuclear fission.

Brainwashing as Psychiatry in Reverse

The centrality of this film to Hunter's new account of brainwashing would be affirmed in testimony he delivered to Congress in 1958, where he cited it as the clearest possible evidence of the Reds' methods and ultimate objectives. Here, too, it is worth noting that among various autobiographical statements whose reliability is hard to determine, he again cited the weird effect of *déjà vu*, the eerie feelings induced by his earliest interviews with subjects of the process. "As I was taking notes, I felt that I had written this all before, and *yet how could I have done so?* I had that same eerie feeling often during that period."

He then suggested that brainwashing could properly be understood only from a medical perspective. There he echoed the concluding passages of his 1956 study. Following its account of Pavlov, subsequent chapters of the book went on to make clear that the blueprint found in the scientist's secret manuscript had provided the red thread linking the Soviet show trials of the 1930s to the case of Cardinal Mindszenty to the US POWs and the "mass hypnosis" of the Chinese population. In each case the pattern was similar: a six-month-long rehearsal on Pavlovian lines prior to the eventual confession or trial, like a course in psychotherapy (or the preparation of a film scenario). Hunter's tendency to put the practice of mind control into a medical framework then reached its logical conclusion: "Red POW camps were simply large clinical laboratories in which prisoners were dealt with as patients and mental cases." Camp personnel used the latest psychiatric methods to achieve mind control; those methods were reverse engineered to achieve something like the opposite of recovery: the creation of a "political neurosis."

Striking here is the projection of elements of the midcentury Western critique of psychiatry onto the scientific practices of the Eastern Bloc. That particular formulation seems to have owed much to Hunter's friend Dr. Leon Freedom, the Baltimore neuropsychiatrist. Freedom was frequently cited in Hunter's 1956 book, which mentioned his expertise in the field of psychiatry and on the political applications of Pavlov's work. Little is known about Freedom, a seemingly marginal figure in the annals of Cold War history. Yet in his capacity as Hunter's guide to the perversions of Red psychiatry, he appears to have been the crucial intermediary in the latter's discovery of Pavlov as the missing link in accounts of the scientific basis of brainwashing. Hunter's archival papers contain several unpublished papers written by Freedom that served as background for his 1956 book, including a chapter based on extended conversations between Hunter and Freedom. Those unpublished writings contain a fuller account of Freedom's view of brainwashing's origins in the materialist doctrines of "cortico-visceral psychiatry." In essence, this was a form of psychiatry in reverse that adapted the methods of modern psychotherapy—free association, the talking cure, the notion of catharsis—to the process of producing "man-made illness" on a mass scale, a "national neurosis." This was the dark side of modern psychiatry, a form of mad science whose further spread posed an extraordinary menace to the West. Even as Operation Big Switch was under way, Hunter wrote, a large conference of psychiatrists, physiologists, psychologists, and biologists had been convened in Peking in September 1953. There they had been treated to lectures on the synthesis of dialectical materialism and Pavlov and received instruction in experimental work on conditioned reflexes. That event had been followed by the creation of a network of labs at institutions across China. The process that now stood revealed to a still disbelieving world was more or less the process represented in Pudovkin's film *The Nervous System/Mechanics of the Brain*. What had been portrayed there for the first time had culminated in the Chinese newsreel of germ warfare confessions. Freedom's description of the footage was remarkable for its language: he described the US servicemen confessing and asking forgiveness "exactly as the hysterical converts of a hillbilly snake healing faith might shout to be 'saved.' . . . The behavior of the officers is hysterical in the extreme." "*1984*," he noted ominously, "is closer than we think."* Even as official reports dismissed brainwashing

---

* Thomas Pynchon's foreword to the 2003 Penguin edition of Orwell's *1984* underscored Pavlov's presence in brainwashing discourse, echoing the extensive thematization of Pavlov's work in his own *Gravity's Rainbow* (1973).

as science fiction, Freedom cited the speculative futures of science fiction as having attained frightening actuality.

The intersection of brainwashing discourse with that of antipsychiatry reached a kind of apotheosis in a set of documents that circulated in US intelligence circles in the mid-1950s. Their authorship is disputed, though certain indicators point to Hunter's friend the psywar expert Paul Linebarger as a likely source. There, too, a selective account of the history of Pavlov's research was woven into the narrative of Communist mind control. Titled "Brainwashing: The Communist Experiment with Mankind," the first took the memoir of one US POW as exemplary of the Communist "application to human beings of Pavlov's conditioned behavior pattern," which, when applied systematically to a person, would make him "as docile as a dog on Pavlov's experimental table." The account added fresh details from Pavlov's life's work, including a turning point that had been reached when, one day in 1924, his laboratory in Leningrad had been flooded during a violent storm. The shock of that occurrence had undone all the conditioning of his dogs' carefully calibrated nervous systems: "Through this accident Pavlov had brought about a nervous breakdown in the animal by external means." Much of the Kremlin's subsequent mind control program could be understood as an extension of the lessons learned during that disaster. In effect, the regime was carrying out an experiment in the artificial production of neurosis through reflex conditioning, achieved by means of words and suggestions. Within the tremendous "scientific prison" that it maintained, the suggestions were conveyed through "an unending stream of manifestos, newspapers, speeches, plays, radio, movies, etc. The words and suggestions become strong engrams [memory traces], work their way into the brain cells, become part of the subject's thinking."* A concluding section titled "Prisoners of War, the Human Experiments" discussed countermeasures

---

* Hunter wrote that in one such play, *The Question of Thought*, the Reds had laid bare their technique, "just as though a brain surgeon were capable not only of exposing the tissues of mind but the thought processes and reactions that go on inside them." Hunter, *Brainwashing: The Story of Men Who Defied It* (New York: Farrar Strauss and Cudahy, 1956), 113.

against the process, stressing the need for knowledge of the enemy and the "perverted science" he employed to achieve his sick ends.

Attached to the document was a pamphlet titled "Brainwashing, A Synthesis of the Russian Textbook on Psycho-politics," along with an accompanying evaluation by a member of the Psychological Warfare Board (again, possibly Paul Linebarger) that advised Eisenhower on psywar. Alleged to be the work of Josef Stalin's henchman Lavrenty Beria, the pamphlet included accounts of the myriad ways in which the methods of contemporary psychiatry—from hypnosis and psychoanalysis to electroshock and psychosurgery—were being applied in the Soviets' no-holds-barred campaign for absolute dominion over the mind. As interesting as the contents of the lurid document was the statement by the author of the accompanying evaluation, which perfectly illustrates the feedback loop between paranoid fact and paranoid fiction within brainwashing discourse. Acknowledging doubts about its provenance, the author nevertheless wrote that "if the booklet is a fake, the author or authors know so much about brainwashing techniques that I would consider them experts, superior to any that I have met to date." In a telling illustration of the way the CIA fell under the spell of its own propaganda, the fact that the pamphlet might be fictitious did nothing to lessen its truth value. As proof, the author referred to what in all likelihood was the Soviet film *The Nervous System*: "As for the work of Pavlov being carried forward by later Russians, this is true and experiments continue. Evidence appears in authentic films now in US Government possession." The evaluator warned the unnamed senator to whom his evaluation was submitted to exercise due caution in the use or release of the pamphlet: "The story is so fantastic that few people would believe or want to believe it."

Conclusion

The fantastical elements to which this author alludes have long been essential features of the brainwashing story. Paradoxically, they may help explain both the staying power of the concept and the air of disrepute that hangs over it. It is a weird artifact of the panicky climate that marked the early days of the Cold War, yet as scholars have newly begun to appreciate, it is an essential springboard to the wider history of that conflict. In recent years, that largely forgotten story has been recovered as part of the prehistory of the torture scandal that grew out of the war on terror. Guantánamo's links to MKUltra and to the CIA's classified interrogation manual *KUBARK Counterintelligence Interrogation* have now been widely reported on. At the same time, the term *brainwashing* has achieved new currency in connection with the phenomenon of online recruitment and radicalization; the uneasiness that now surrounds the dark potential of modern social media mirrors the anxiety that surrounded midcentury mass media.

In this regard, it is not surprising that Hunter placed such emphasis on the obscure Soviet film *The Nervous System* in his account, for brainwashing was, among other things, a function of modern communications theory, a field that came of age in the 1950s (see chapter 5). The fascination and unease that surrounded new media and their effects on self and society were important preconditions of brainwashing's emergence in the early Cold War, particularly as, under the influence of new theories of perception, learning, and conditioning, those effects were more closely linked to the mechanisms of mind and brain. "Mind control," in other words, belonged to the spectrum of effects that, according to communications theorists such as Marshall McLuhan and others, arose out of the nexus between modern mass media and the human nervous system. Edward Hunter's discovery of *The Nervous System* and the paranoid narrative he wove around it belongs squarely within this nexus.

Could the mass media be used to transform men's basic thought patterns, implant new ones, condition new loyalties? Could a skilled interrogator accomplish those aims? If so, at what level of the self? Did means exist to seize possession of the "nervous patterns" of human behavior? For many people at the dawn of the Cold War era those were not simply idle questions, though few pursued answers with the same degree of literal-mindedness as Hunter. That literalism has often been regarded by skeptics as a central fallacy in the concept of brainwashing. As the CIA manual *KUBARK* stated plainly on its first page, there is nothing mysterious about interrogation, and subsequent scholarship has tended to confirm that judgment. In its fixation on the *scientific* origins of torture, Darius Rejali suggested, brainwashing discourse betrays a modern theodicy, a "peculiar modern conceit about where evil comes from in our age." Yet although the notion of brainwashing as an esoteric form of modern science may well, as critics argued, have been a phantasm, a myth, it was, so to speak, a concrete myth, one with all-too-real effects. Edward Hunter did not simply pluck the elements of this construct out of thin air; rather, in an almost mediumistic fashion, he channeled broader scientific and cultural currents and the new possibilities and dark forebodings they opened up. For all the fictive elements of his account, it cannot be dismissed as simply disinformation, for to do so would be to ignore how deeply embedded it was within the wider currents of the time—currents in which science, fiction, and speculative knowledge, or science fiction, mixed freely. Fantastical as it was, Hunter's account emerged as a natural possibility out of the constellation of developments that marked the simultaneous birth of midcentury brain science and the onset of the Cold War. That constellation created conditions in which it became possible to conceive of humans as almost infinitely malleable, in which human faculties such as perception, knowledge, and memory became "substrates" to be worked on, whether in a laboratory setting or in the political realm. To chalk Hunter's account up merely to myth is to underestimate the role of myth as a historical force in its own right.

It is telling in this regard that Hunter's personal correspondence with Leon Freedom includes a citation from the neurophysiologist Grey Walter's book *The Living Brain* (1953), whose work I explored in chapter 1. In a brief note written in 1954, he drew his friend's attention to a passage in Walter's book: "No mental theory or practice is likely to survive which does not take into account the principles of cerebral functions revealed by physiology any more than the practice of medicine can ignore other physical functions. Already the new type of psychiatrist, in touch with centers of physiological research, has adopted a new outlook." Hunter thus identified Walter as a member of a new scientific vanguard, a role that he had assumed in the early 1950s along with Wilder Penfield and others whose findings had begun to filter into broader public consciousness. It is their spirit that hovers over the first chapter of Hunter's 1951 book.

But the scientist Hunter most directly dwelled on in his published writings is Pavlov, who was central to the new account set forth in his 1956 book. Hunter was not alone in paying homage to Pavlov; many of those who contributed to Cold War discourse wove variations on the role that the Russian physiologist's findings had played in providing the Kremlin with the tools it needed to make good on its promise of creating a "new Soviet man."* Yet in that regard it was parasitical on a much broader scientific literature of the period that extended from neuropsychology and psychotherapy to communications theory and public relations. In all those fields the work of Pavlov was foundational. As Grey Walter himself showed, midcentury research into learning, formerly limited to the mind, began under the influence of Pavlov's findings to be extended into research on brain processes and neural pathways.

So, too, did breakdowns or pathologies of the learning process take on new meaning in relation to changing understandings of the brain. Under conditions of total "milieu control" such as a POW camp, the mind could be induced to experience vivid distortions of perception and memory, even to the extent of coming to believe in its guilt for an alleged crime and, most bizarrely, supplying the memories that provided confirmation of it. Perhaps it was the discovery of the extent to which even the seemingly most steadfast mind may play tricks on itself that conferred on brainwashing its uncanny nature. As we shall see in subsequent chapters, that discovery grew all the more unsettling as fears of brainwashing were turned against American society. Increasingly the haunting power of this notion came to reside in the possibility that the enemy was not simply "out there" but rather operated within society's very midst. The groundwork had already been laid within antipsychiatric discourse of the early 1950s, a discourse that found many echoes in Hunter's work, with its portrayal

---

* The Polish poet Czesław Miłosz noted of Pavlov that although he was a very religious man, his theory of the conditioned reflex constituted "one of the strongest arguments against the existence of some sort of constant called 'human nature.'" Miłosz wrote that his own revolt against the Soviet system came not from his "reasoning mind" but from his stomach, i.e., his "nature." Miłosz, The Captive Mind (New York: Knopf, 1953), 206–7, xii–xiii.

of the brain—now stripped of its status as a "divine creation"—falling
subject to the influence of powerful forces of many kinds, not least those
associated with the modern mass media. It is striking the extent to which
Edward Hunter's writings seem to reproduce many of the symptoms that
he associates with brainwashing. We may in the end wonder whether
the "eerie feelings" to which Hunter's writings always returned are an
expression of the question: Was the enemy a distorted mirror reflection
of ourselves?

HAYES'S HALLUCINATION

The first thing a human being is loyal to . . .
is his own conditioned nervous system.

—YEN LO, CHINESE BRAINWASHING EXPERT
(RICHARD CONDON, *THE MANCHURIAN CANDIDATE*)

Throughout the 1950s, as we saw in part I of this book, Western scientists conducted groundbreaking "experiments in consciousness." So, too, Western cold warriors often argued, the Soviets and their allies were engaged in carrying out similar experiments, both on their own citizens and on foreign nationals such as the US servicemen captured during the Korean War. "It is man's consciousness," proclaimed one memo that circulated in US intelligence circles, "with which the Communists experimented on US POWs." In the wake of Operation Big Switch, the repatriation of US POWs in fall 1953, debates surrounding the putative results of those experiments became the subject of a voluminous literature. They also became part of the testimony submitted in court-martial trials. In the case of former POWs forced to explain statements made in captivity or their initial refusal to be repatriated, brainwashing became a central element in the defense strategy crafted by lawyers and medical experts. Faced with Red mastery of the dark arts of mind control, those experts argued, how could unworldly, poorly trained young Americans from small towns in Texas or Kentucky be expected to remain steadfast in their beliefs?

That was the approach adopted by experts such as Columbia University psychiatrist Joost Meerloo and Edward Hunter's friend Dr. Leon Freedom in the cases of two servicemen, Claude Batchelor and Frank Schwable. The accusations against Schwable and Batchelor were serious, including charges that the two men had made radio broadcasts on behalf of the enemy in which they had condemned capitalism, the Americans' use of biological warfare against North Korea, and domestic injustices ranging from the Jim Crow laws and the vigilante actions of the KKK to the executions of Julius and Ethel Rosenberg. The testimony of the two experts included an extensive discussion of the role played by Pavlovian methods in the process of conditioning that culminated in brainwashing. (Schwable somewhat disturbingly also compared the process to the sensory bombardment of American advertising.)

According to Dr. Freedom, testifying on Batchelor's behalf, the fact that the accused could not tell right from wrong was due to a "mental disorder in the 'political field,' an 'induced political psychosis'" that he compared to schizophrenia. In a manner highly reminiscent of the discourse of midcentury antipsychiatry, Freedom argued that the men had been victims of such monstrous abuses of clinical expertise that they could in no way be held responsible for their failure to resist. Edward Hunter's correspondence with Freedom and with Batchelor's attorney, Joel Westbrook, repeatedly touched on many of the same points. They are echoed in a lengthy letter Westbrook sent to Batchelor's mother in which he described the strategy he had adopted in concert with Freedom and referred, among various forms of scientific evidence he planned to submit as affidavits, including accounts written by former prisoners of Hungary's Communist regime, to Pudovkin's film on the work of

Pavlov. The second reel of the film, he explained, showed the telltale experiments on human beings that exposed Communist intentions in their most naked form. In the end, however, the court remained unpersuaded and ruled against the legitimacy of the brainwashing defense.

What actually happened, or was believed to happen, in brainwashing? In their efforts to describe the process, accounts of the period often referred to literary portrayals such as those in George Orwell's *1984* (1949) or Arthur Koestler's *Darkness at Noon* (1940). Koestler's novel was a frequently cited touchpoint in the Cold War literature on Communist mind control. In its portrayal of the dark side of the Soviet regime's efforts to transform human consciousness, and thus the course of history, it was treated as having near-documentary status.* It showed how probing the inside of people's heads, the better to understand their workings and how best to mold them in accordance with the revolution's avowed ideals, had become an abiding preoccupation for the Bolsheviks—particularly as questions of orthodoxy and deviations from it took on an increasingly lethal character. Written in the aftermath of the bloody show trials of the mid-1930s, it centers on the figure of Rubashov, a Communist Party official who once dedicated himself to rooting out so-called enemies of the people but now stands accused of being one of those enemies. From the prison cell where he is kept while undergoing repeated interrogation, he records in his diary the following statement about the attitude that prevailed within the Party: "We admitted no private sphere, not even inside a man's skull" (a line later echoed in Orwell's *1984*).

It is in keeping with this that Rubashov passes the time in prison by speculating about the insides of the skull of the Party leader, a figure modeled on Stalin: "What went on in No. 1's brain?" He pictured to himself a cross section through that brain: "The whorls of grey matter swelled to entrails, they curled round one another like muscular

---

* The Bolshevik project had earlier been satirized in Mikhail Bulgakov's novella *Heart of a Dog* (1925), the story of a dog transformed into a man in a laboratory experiment involving a brain transplant.

snakes, became vague and misty like the spiral nebulae on astronom-
ical charts. . . . What went on in the inflated grey whorls?" He ends up
fantasizing about a classroom of the future, where a scientific form of
history is taught by instructors who illustrate their lessons with images
of No. 1's brain, pointing out to their pupils the "grey foggy landscape
between the second and third lobe."

The flip side of the Soviets' cult of No. 1's brain was, as Rubashov's
situation illustrates, the extreme vulnerability of that organ to the
techniques brought to bear on it by the regime's interrogators. His
own diary now becomes a central exhibit in the case being assembled
against him in the weeks leading up to his final confession. In this
system it is not enough simply to convict a person of crimes against
the state; the trick is to extract a confession to those usually completely
fabricated crimes. In the end it is the ritualistic performance of transformed consciousness—confessions and staged trials (which are then
followed by executions)—that matters more to the regime than sincere
belief in Party doctrine. Anticipating the fate that awaits him following
his confession, Rubashov notes bitterly that "the brains of the revolution's leaders are now filled with lead."

Edward Hunter's own quasi-literary attempts to capture the mystery of
what occurred in the POW camps included an extensive account of the
ordeal undergone by one John B. Hayes. That brainwashing was comparable to one of the altered or dissociative states that so fascinated
midcentury scientists became clear in his treatment of Hayes's case in
a section of his book *Brainwashing: The Story of Men Who Defied It*. The
section is titled simply "Hallucination." Hayes had worked for many
years as a missionary in China before being arrested and subjected to
intensive interrogation on charges that his missionary work had served
as cover for his role as the head of a spy network. At the end of an exhausting nine-month-long period of "suggestive interrogation," he
finally broke down.

"They were chiseling away at my memory," he later reported to
Hunter. Taking care to mention the fact that Hayes was familiar with
the ideas of Pavlov, Hunter described the climax of the process in vivid

terms. It took the form, in Hayes's words, of a flashback that jolted him like "a lightning stab of memory." "Like a bolt of lightning," Hunter wrote. "The scene came back to him as if it had happened that same day. How could he have forgotten it? He saw it all now in his mind's eye, all over again, exactly as it had happened. Indeed, how could he have forgotten!" The incriminating scene rose up before his mind's eye in what he took to be his memory: "The time, that is, when his friend came to him in his house while he was still only under detention and they chatted and this man remarked in a worried way, 'By jove, I better get rid of that transmitter!' He heard the words distinctly."

Confronted by the seemingly irrefutable evidence of this scene, Hayes was ultimately powerless to resist giving his captors the confession they wanted. This moment was accompanied by a deep, almost pleasurable feeling of release—an experience that Hunter elsewhere likened to the catharsis experienced at the conclusion of a process of talk therapy. The hallucination was nearly cinematic in its details: "Under the uninterrupted demands made upon his mind in that grotesque environment . . . Hayes had attained a clarity very much like a hypnotist's subject, who can recreate from deep within his subconscious some exact memory of a long past incident which he had believed gone entirely from him." It was just as scientists such as Wilder Penfield had theorized, except that in Hayes's case the "flashback" was a completely false record, the result of an arduous process of conditioning and suggestion.*

Yet even as Hayes became convinced of the authenticity of his memory, he retained a faint awareness of his circumstances. Here again his experience can be related to another of the strange findings of midcentury brain science, for in this respect his experience accorded with Penfield's findings about "hallucinative reminiscences." Hayes's trancelike state was, in fact, a strange kind of "double consciousness": "I

---

* This scene bears some resemblance to a description of the syndrome known as the "prisoner's cinema," in which sensory deprivation produces hallucinations. See Ronald Siegel, *Fire in the Brain: Clinical Tales of Hallucination* (New York: Dutton, 1992), 211–28; Oliver Sacks, *Hallucinations* (New York: Vintage Books, 2012), 34–44.

believed," he stated, "I had had a hallucination and I believed I hadn't."
Hunter seized on that as evidence of some surviving core of resistance,
a hidden reserve of the self into which the conditioning process could
not penetrate: "Is that what psychiatrists call ambivalence, the division
of the brain into separate compartments?" If so, perhaps it contained
grounds for hope: "If truth can linger in the mind despite the strongest
of hallucinations, . . . the reason is clear why the Reds [can] . . . never
capture their minds completely!"

CHAPTER 4

The "Prisoner's Cinema"

Interrogation in the Age of the Explorable Brain

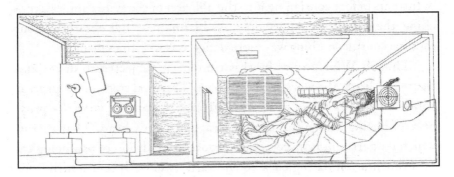

The isolation chamber in Donald Hebb's lab at McGill University. In addition to wearing vision- and noise-masking devices, subjects had EEG leads attached to their heads to monitor their brain activity.

One of the most famously disturbing scenes in the literature of totalitarianism occurs near the end of George Orwell's *1984* (1949). The novel's protagonist, Winston Smith, following his arrest on charges of belonging to the underground opposition, has undergone the interrogation to which all prisoners are subjected in order to extract confessions from them. He now stands at the brink of a complete breakdown. His memories are scrambled, discontinuous. Amid his mounting confusion fragmentary scenes flash before his eyes. He seems to remember confessing to assassination, sabotage, and other crimes. Hallucinatory images overwhelm him, cinematic in their intensity: "He was rolling down a mighty corridor . . . roaring with laughter and shouting out confessions at the top of his voice." Smith now enters the final stage of his breakdown. At the end of the process he will be fully conditioned to accept his own confession.

An essential feature of the process involves its repeated efforts to chisel away at and finally destroy what his interrogator, O'Brien, calls his "defective memory": in particular, his conviction that in his position at the Ministry of Truth, he had seen evidence proving that the confessions of former Party members were false. O'Brien states, "There was a certain photograph about which you had a hallucination." The photograph does indeed exist and O'Brien shows it to Winston, but only for the briefest of instants before consigning it to the oblivion of the state's amnesia machine, the so-called memory hole. As Hannah Arendt observed about totalitarian rule, nothing was more dangerous to it than memory.

Though Winston has by this time been subjected to drugs and electrical shock, the worst is yet to come. The final stage in the process comes when he's taken to Room 101 and threatened with the prospect of having a cage full of rats strapped to his head. That threat, which renders him fully compliant, evokes a tradition of psychological experimentation dating back to the early twentieth century. By contrast with the previous chapter, which focused on Western speculation about Ivan Pavlov's contributions to the Soviets' campaign of ideological conversion, this scene points to an American lineage in the history of behavioral conditioning: the work of John Watson. Partly inspired by Pavlov, Watson launched a research program centering on rats and mazes (before abandoning it for a career in advertising). It was that tradition, which became dominant in much Western psychology, that Orwell invoked, specifically Watson's experiments with fear conditioning of human infants, using in this instance not rats but white rabbits. (Like Pavlov's work, it was captured in a film, of Watson's subject Little Albert, made in 1920.) Midcentury dystopias of behavioral control and its extension to torture were based as much on Anglo-American as on Soviet paradigms. As Darius Rejali has observed, amnesia also plagues the Western consciousness of the relationship between torture and democracy.

During the Cold War, Orwell's novel became a touchstone of accounts of the dangers facing the West. The assault on memory associated with Red mind control achieved near-iconic status in this novel, and its representation of that threat became central to the way the Cold War was framed.* Many people were prepared to take seriously O'Brien's threat to Winston that "We make the brain perfect before we blow it out"—perfect, that is to say, in the sense of wiped clean of memory. The possibilities envisioned by Orwell acquired a more than merely fictional status; they were treated as having evidentiary value, both diagnostic and predictive. As Edward Hunter's friend Leon Freedom put it, "1984 is closer than we think."

This brief account of the spell cast by Orwell's novel over the Cold War imagination serves as the point of entry into an examination of the United States' own mind control program. Tracing the work of the neurologist Harold Wolff, one of the figures entrusted by Dulles with overseeing that program, before turning to that of Donald Hebb, the Canadian psychologist whose pioneering research into sensory deprivation became a central element of that program, this chapter concludes by briefly considering the denouement of the "Dulles era" in *KUBARK*, the classified interrogation manual finalized by the CIA in 1963.

1984, as this chapter illustrates, was closer in more ways than one. It was not simply that the future now seemed terrifyingly actual; whether mind control was conceived as acting on individuals or on entire publics, the fact was that the process had a demonstrably homegrown dimension. During the 1950s, mind control American style acquired its own reality. As we saw in the previous chapter, Orwell's representation of the role of

---

* If *1984* portrays the destruction of memory as the essence of totalitarianism, stories such as that of the Korean War POW General William Dean celebrate the near-superhuman powers of recall (including of the number of flies he killed while in captivity), which enable resistance to mind control. See William F. Dean, *General Dean's Story* (New York: Viking, 1954).

the media in the totalitarian state was mirrored by the role taken on by those media in managing the Western public's perceptions, emotions, and memory during the Cold War. Thus the adaptation of his novel for the screen in 1956 was undertaken in a production partly sponsored by the CIA.

Or consider the film *POW* (1954). That Ronald Reagan vehicle was made to help counteract the damage inflicted by the Chinese propaganda coup of 1952, when US servicemen confessed to germ warfare in a media spectacle transmitted around the globe. In the film, American POWs confess to war crimes to their captors, but only in a charade to earn their trust so they can later expose the abuses occurring in the camps. The confessions are false, in other words, not because they are coerced but because they are play-acted. While the Russian camp commandant lectures about the "conditioned reflexes of the rat," the prisoners, forced to watch footage of confessions, react by destroying the film projector. *POW*'s production initially received support from the Department of Defense, but in the end the entanglements of coerced and staged confessions became so confusing that the department withdrew its support.

Such was the strange epistemology of a moment in which the mystery of brainwashing, coupled with the secrecy doctrine adopted by the national security state, engendered a deep crisis of public knowledge. That crisis and the resultant uncertainty about how we know whether what we know is true has haunted Cold War history up to the present. On the one hand, mind control has become what one author has called "the world's favorite conspiracy theory." On the other, the all-too-real elements of the story remained half forgotten for decades, until the torture scandal that arose out of the war on terror forced them back into public consciousness in the early 2000s. One of many reminders of the ways the Cold War refuses to end has come with the rediscovery of the United States' own interrogation or mind control program dating back to the 1950s. With that rediscovery we have come full circle: if Winston Smith's final breakdown is brought about by exploiting his fear of rats, the 2015 report released by the APA in response to the torture scandal brought to light evidence that American psychologists employed by the

CIA had exploited medical data regarding detainees' phobias as part of the US interrogation program. In that program, clinical knowledge was reverse engineered for the purpose of torture. It is, Rejali has observed, one of the comforting illusions of democracy that torture is practiced only elsewhere.

Mind Control in the Age of the Explorable Brain

Amid the circumstances described above it is hardly surprising that a novel such as *1984* could attain the status of a truthful representation. If brainwashing, as some have argued, was a "strategic fiction," a means of sharpening Americans' focus on the nebulous conflict at the heart of the Cold War, then fiction logically held a significant claim to knowledge of it. Yet this situation was not without its absurd features. By the mid-1950s, so deeply entangled had paranoid fact and paranoid fiction become, and so confusing were the consequences, that at a symposium on the topic of forced indoctrination held in 1956 by the Group for the Advancement of Psychiatry, its presiding chair, the neurophysiologist John C. Lilly, felt obliged to remind participants of the need to "clarify the differences between the fantastic account of Orwell and the real processes actually used in authentic cases." A comprehensive report on the phenomenon of Communist indoctrination and interrogation was presented to that body. Its authors, Harold G. Wolff and Lawrence E. Hinkle, sought to dispel the many myths encapsulated in the term *brainwashing* and to provide a more realistic account of the phenomenon.

Yet this process of demystification had its own decidedly ambiguous aspect. On the one hand, as we saw in the last chapter, the notion of brainwashing, in all its lurid details, continued to cast a powerful spell over the highest levels of the US government, while at the same time taking on a life of its own in the public realm, where it became a staple of social critique, antipsychiatric discourse, and pop cultural representation (see chapter 5). On the other hand, many of the same figures who pursued that debunking operation undertook research programs quite as fantastical as those attributed to the enemy. Harold Wolff himself occupied a pivotal position within a clandestine enterprise of such vast dimensions that the historian Alfred McCoy has dubbed it a "Manhattan Project of the mind."

It is all the more ironic that that research formed part of a wider program founded on the tantalizing promise of a total, fully integrated science of the self. That promise was embodied in the blandly named Society for the Investigation of Human Ecology (SIHE), which helped channel government funding to a small army of researchers in psychiatry, psychology, anthropology, and allied sciences. At the center of that web stood the relationship between Allen Welsh Dulles and Harold Wolff, an eminent neurologist who had become close to the CIA director while treating his son following the brain injury he had suffered in Korea (see Clinical Tale 3). Shortly after his appointment as spy chief in February 1953, Dulles authorized funding for MKUltra, a classified program of research in behavioral science with a remit to carry out wide-ranging experimentation in the field of "brain warfare." That research continued for the rest of the Dulles era at the agency, culminating in the classified interrogation manual *KUBARK*. Since its shadowy origins in the 1950s, MKUltra has entered the annals of Cold War lore as a byword for science bereft of ethical restraint, particularly in connection with the work of Ewen Cameron, a figure whom Naomi Klein, Alfred McCoy, and others have placed at the center of an overarching narrative of US history in the Cold War era, though other scholars have challenged aspects of that account.

Though MKUltra had its precursors, under Dulles the US intelligence community's research into mind control became funded and systematized on an entirely new scale. The point person for much of this was Harold Wolff, the head of SIHE. Under Wolff funding was approved for a host of research programs operating at the frontiers of the sciences of brain, behavior, and mind. Few areas were left unexplored. Though no reliable science of mind control ever emerged from the research, the work yielded many unexpected fruits. As the author John Marks put it, CIA funding for young and often iconoclastic researchers opened up the field of the human sciences and helped break the behaviorist monopoly over many of those disciplines. Indeed, he suggested, pointing to continuities with some of the cultural practices of the 1960s, research in the human sciences had both a mind control side and a transcendental side. The well-known story of the CIA's LSD research is one part of that. Another is sensory deprivation, Donald Hebb's brainchild. In the hands of John Lilly, Cold War research into sensory deprivation found its way into *KUBARK*, while also later sowing many of the seeds of the counterculture.

The 120-plus-page report issued by Hinkle and Wolff is fascinating on many levels, not least because it shows just how seriously such figures took brainwashing (or, as they preferred to call it, "interrogation and indoctrination")—not simply as an element of Cold War strategy but as an exemplary instance of midcentury techniques of getting inside people's heads. Even if, as they were at pains to demonstrate, Communist interrogation was not based on scientific methodology, the effort to understand it demanded a scientific approach. Apart from their interviews with former prisoners and their reliance on classified and published reports, their analysis therefore drew upon the results of what they discreetly referred to as "laboratory and clinical investigations . . . carried out in order to throw light on the psychological and physiological processes involved" in interrogation. Though the practice of interrogation did not involve any formal training in psychiatry, psychology, or pharmacology (all of which were speculated about in connection with the US POWs), its effects could be reproduced in experiments showing how easy it was to generate a "situation of frustration" within a subject.

On the face of it, the procedure was simplicity itself: no scalpels, drugs, or other esoteric methods were involved, just the age-old ones of isolation, sleep and food deprivation, and temperature manipulation. Over time their combined effects led inexorably to profound disorganization of the prisoner's personality. Symptoms included nightmares and visual hallucinations (e.g., of God or the prisoner's wife). Eventually the prisoner came to doubt his own memory, becoming suggestible to the point where he was ready "to confabulate the details of any story suggested to him." Once the preparatory phase was complete, the crucial scenes were acted out in the interrogation room: "The lighting is arranged so that the prisoner can be placed in a bright light, while the interrogator sits in relative darkness. Sometimes a stenographer is present in one corner of the room to take notes." The scene that followed was a form of talk therapy in which the interrogator encouraged the prisoner to talk at seemingly interminable length. Through a drawn-out process of repetition, correction, and suggestion, he guided the prisoner through and eventually out of the "maze in which . . . [he] finds himself." The invariable result was confession. The question was frequently posed: Why did prisoners confess to fake crimes during public trials? The answer offered by Wolff and Hinkle was reassuringly simple: Though speculation surrounding the phenomenon was understandable, there was simply no basis for the view that the Communists used "occult" methods. Nor were there grounds to believe that the methods used in such cases produced lasting changes in brain function or memory; the alteration of belief was purely temporary.

Yet Wolff and Hinkle's demystification of brainwashing, persuasive as it was, rang hollow. As the conclusion noted, though no science was involved, the Communist procedure nevertheless mirrored many of the methods of modern medicine; though the Communist management of prisoners was not designed or carried out by psychiatrists or neurophysiologists, "nevertheless the interrogation does deal with the prisoner by using many of the same methods that the physician uses in the management of the patient." And in a strangely circular form of logic, Wolff compared the methods he used in treating his patients to those of an interrogator.

"No Day Without Its Experiment"

At the time of his recruitment into MKUltra, Harold Wolff was one of the world's foremost neurologists, serving as head of his department at Cornell University Hospital in New York City from 1932 to 1962. He grew up in modest circumstances, graduating from New York's City College in 1918 before entering Harvard to study medicine and then Johns Hopkins to study psychiatry. His early training included a period of study in the Leningrad laboratory of Ivan Pavlov. A recent biographical sketch suggests (probably apocryphally) that his stay in Leningrad coincided with a major flood that engulfed the city in 1924. Apart from the damage it wreaked on the city itself, the flood had a major impact on the course of modern physiology, for the trauma of the raging waters that found their way into Pavlov's laboratory seemed at a stroke to undo all the meticulous conditioning of his dogs' reflexes and nervous systems. Confronted with this development, Pavlov subsequently devoted much of his research to investigating the implications of that so-called catastrophic reaction for conditions such as war neurosis. In the 1950s, that moment was emphasized by brainwashing authorities such as the British psychiatrist William Sargant, who pointed to the Leningrad flood as an event illustrating how the brain might be "wiped almost clean . . . of all the conditioned behavioral patterns imprinted in it."

Whether or not Wolff was present for that dramatic event, he remained a lifelong disciple of Pavlov. Though the report he coauthored rejected charges that a grand Pavlovian strategy lay behind Communist successes in the field of interrogation, many of his own investigations into this field were inspired by Pavlov's findings concerning "failures of conditioning." Wolff maintained a frenetic pace in his research, which he oversaw from an office featuring a sign that proudly proclaimed "No day without its experiment."* His major focus was migraine, a condition from which he himself suffered and on which he became a leading authority. Just as Grey Walter's and Wilder Penfield's contributions to the midcentury revolution of brain science arose out of their research on epilepsy, so, too, Wolff's research on migraine—a condition that shares certain features with epilepsy, notably the aura and hallucinations that accompany some forms of it—galvanized advances in the field. Wolff

---

* By the mid-1950s, his work in such areas as stress, migraine, and hypnosis had brought him much public notice. His personal papers include a letter from one man who wrote to propose that he conduct an experiment in which he would hypnotize a doctor engaged in cancer research and then ask him to write out a formula for the cure of cancer that he otherwise had no conscious knowledge of. Wolff responded by politely thanking the man for his "interesting comments." (James S. Adams to Wolff, May 3, 1957, Miscellaneous Correspondence, Wolff Papers.)

viewed migraine as a condition rooted in a serious "disturbance of man's total relation to his environment." Conceived in that way, the treatment of the condition demanded an approach that integrated disciplines ranging from neurology and psychiatry to social anthropology. In that approach were contained the seeds of the Society for the Investigation of Human Ecology (SIHE). Wolff's efforts to find treatment for the migraine-afflicted included a wide range of unorthodox methods, among them a human centrifuge. They also drew on the newly emergent field of research into sensory deprivation pioneered by Donald Hebb. Like a Communist interrogator faced with a recalcitrant prisoner, Wolff sought to overcome his patients' reluctance to change their ingrained habits. As Rebecca Lemov has written: "In an effort to make his patients more receptive to his suggestions, Wolff tried sensory deprivation chambers where he kept headache sufferers locked in until they showed 'an increased desire to talk and to escape from the procedure.'"

Under Wolff the SIHE became a major conduit for channeling funds to researchers. The organization's first annual report, issued in 1957, laid out a comprehensive plan of investigation on several fronts. The plan was conceived of as being interdisciplinary in the widest sense, drawing on the findings of researchers in a strikingly diverse array of fields whose collective expertise lent itself to the project of fashioning a truly "ecological" paradigm. The need for that arose out of what Wolff termed the "changes occurring in an age of rapid technological advance." The results of Communist indoctrination were but one illustration of the discovery that, in the present world, "the human personality is not as stable as we often assume"—a statement that could serve as an epigram for the entire era. In accordance with that fundamental premise the report outlined several projects ranging from a POW study to studies at McGill University led by Ewen Cameron on "psychic driving" (on which more later), LSD research, and, following the failed Hungarian Uprising in 1956, a study of political refugees from that country. For an ambitious scientist anxious to please his new masters, the mid-twentieth-century "age of extremes" provided unparalleled opportunities for research into human behavior.

It was only natural that brain science would assume pride of place within that interdisciplinary approach, given the brain's status as what Wolff called the "master organ," the site in which the highest integrative functions in the human organism take place. That was already implicit in the commission he had received from his patron Dulles, who had authorized him to pursue research on the brain and nervous system as factors in war, building on his own earlier studies of combat stress. Thus Wolff headed up projects such as Studies of Impairment of Highest Level Brain Functions Following Prolonged Stress.* He also pursued research on "experimentally produced sleep" and "experimental nervous breakdowns."

Another study, Effects of Chemical Agents on Bodily Functions, Mentation, Attitude, Etc., marshaled the combined efforts of a team of neurologists, neurosurgeons, psychiatrists, and psychologists at New York Hospital. The report described the work as pathbreaking; it aimed to shed new light on the highest integrative functions of the brain by studying the mechanisms involved in thought, emotion, behavioral conditioning, and memory. The report implied a willingness to conduct tests on human subjects, including both those with "normal brains" and those with psychological disorders and various forms of "brain damage," and stated ominously, "Where any of these studies involve potential harm to the subject, we expect the Agency to make available suitable subjects and a proper place for the performance of necessary experiments."

Experiments in Hypnosis

Another of the projects identified in the report was outlined under the innocuous-sounding title "Factors Affecting Behavior, Mentation,

---

* The study built on the work of Hans Selye, which made stress one of the major new clinical paradigms of that era. The conceptual framework on which Selye and Wolff based their ideas about adaptation to stress was partly drawn from Grey Walter's cybernetics. Mark Jackson, *The Age of Stress: Science and the Search for Stability* (New York: Oxford University Press, 2016), 138, 211. For an exchange between Wolff and Walter on stress in war and POW camps, see "Psychosomatic Illness," Folder 15-1, Wolff Papers.

Attitude, Etc." The document enumerated a wide range of methods for "influencing" an individual or conversely preventing such influence. The notion of influence was framed as encompassing both familiar social and cultural processes (child rearing, education, military training) and so-called special procedures ranging from psychotherapy and salesmanship to deprivation, hypnosis, coercion, and torture. To assist him in his research, Wolff requested all existing agency documentation relevant to the topic of interrogation, including the following methods: "coercion, imprisonment, isolation, deprivation, humiliation, torture, 'brainwashing,' 'black psychiatry,' hypnosis and combinations of these with or without chemical agents."

The proposal went on to list the questions that investigators would seek answers to with respect to hypnosis (some words redacted): Can ___ be hypnotized? What percentage of ___ are susceptible to hypnosis? How complete is posthypnotic amnesia and its degree of permanency? Can an individual be made to perform acts contrary to his conscious will? Can hypnotic influence be detected? What is the effectiveness of chemical agents in hypnotic procedure? In its conclusion the report promised that once those questions had been properly investigated, it would be possible to assess the effectiveness of hypnosis as both an offensive and a defensive intelligence weapon.

Here we confront a core element of the American response to brainwashing and the panic it engendered—a panic, it must be kept in mind, that had in many ways been produced by the CIA itself. As this and other reports make clear, hypnosis loomed very large in the imagination of US cold warriors. Few if any other methods enjoyed the same aura of mystery and transgressive possibility. The experiments alluded to in the report formed a central motif of the way brainwashing was imagined by figures such as Edward Hunter and represented in popular culture by films such as *The Manchurian Candidate*, the story of a hypnoprogrammed assassin. But they were not just part of the lurid mythology surrounding Communist mind control, for they had a "solidly academic pedigree." Indeed, they built on a tradition of psychological experimentation that dates back to the nineteenth century and is in many ways coterminous with the history of the modern discipline of scientific psychology.

The specter of "hypnotic crime" was first extensively investigated by psychiatrists and experimental psychologists at the turn of the twentieth century and then explored in many variations over the course of the century. In their most basic form, those investigations created scenarios designed to test the possibility that subjects in a trance state could be compelled to commit crimes using weapons that, unbeknown to the subject, were fake: guns loaded with blanks, paper daggers, and bogus arsenic. Such scenarios soon spilled from the laboratory or clinical setting into the public realm, becoming an enduring cultural fantasy in novels and in films such as the German classic *The Cabinet of Dr. Caligari* (1919). They were also evoked at moments of high social or political drama. Such was the case for many of the precursors cited in the literature on brainwashing, among them the Soviet show trials of 1936–1937. At the dawn of the Cold War, that also became true of the notorious trial of Cardinal Mindszenty in 1949. The behavior of the Hungarian prelate, known for his fierce anticommunism, publicly confessing to crimes against the state, was so remarkable that it aroused speculation that he had been placed in a hypnotic trance. The so-called Mindszenty look—a dramatic blank stare later observed among POWs—became one of the hallmarks of the victim of mind control. Coming in the same year as Orwell's *1984*, a novel that repeatedly invoked the notion of hypnotic control, the case of Mindszenty and his alleged hypnotic conditioning became the prototype of mind control. As *1984* illustrated, the cult of personality central to totalitarian rule was often imagined as operating through a system of highly personalized "magnetic" power.*

It was against that backdrop that Cold War scientists conducted their research into hypnosis. With allowances made for the different contexts, the research questions outlined in the Cornell report might

---

* One of the arguments employed by Adolf Eichmann's lawyer in his client's defense was that he had been the victim of a form of "political hypnosis" emanating from the Führer.

have been taken directly from the notebooks of a nineteenth-century experimenter. It remains a matter of speculation, however, how far Cold War experiments of that type were taken. Though many documents of the period have been declassified, the record is incomplete, full of gaps and redactions. In his still unsurpassed account *The Search for the "Manchurian Candidate"* (1979), written in the wake of congressional investigations into CIA scandals in the mid-1970s, the former State Department official John Marks showed that as the phantasm of Red mind control seized hold of the Western imagination, the possibility of hypnoprogramming became a topic of intense fascination. Fictional accounts such as Orwell's played a role in both fueling that fascination and "authenticating" what remained, in the end, a matter of speculation. Such possibilities were then further outlined by figures such as the Yale psychologist Irving L. Janis, the author of a report written for the RAND Corporation in 1949. Titled "Are the Cominform Countries Using Hypnotic Techniques to Elicit Confessions in Public Trials?" it canvassed the existing literature on "hypnotic crime" in connection with the Mindszenty case.[*]

Even before MKUltra received authorization, research into the hypnotic conditioning of agents and the effects of hypnosis in interrogation was being carried out by the CIA. Wolff's research program simply picked up where those earlier investigations had left off; they themselves had posed variations on the same questions that researchers dating back to the *fin de siècle* had posed: Can a person be made to perform acts contrary to his conscious will? To be programmed to commit real crimes or, conversely, confess to fabricated crimes?

Thus we confront the paradox created by the fact that Wolff and his allies rejected the term *brainwashing* and blamed its hold over the popular

---

[*] Janis cited the work of the Russian neuropsychologist A. R. Luria, who in the 1930s had investigated whether hypnotically implanted false memories could be used both to convince a subject that he had committed crimes and to instill the appropriate feelings of guilt. Irving L. Janis, "Are the Cominform Countries Using Hypnotic Techniques to Elicit Confessions in Public Trials?," RAND Corporation, April 25, 1949, https://www.rand.org/content /dam/rand/pubs/research_memoranda/2006/RM161.pdf.

mind on fictional creations, while in their own research they practiced brainwashing in everything but name and did so in ways that mimicked fictional scenarios. Setting their skepticism aside, they seized the opportunity to perform all manner of experiments, as well as the license to go as far as needed. In the end it seems likely that the answers to the questions posed by Wolff were negative. Yet that did little to lessen interest in hypnosis. Indeed, some of the most intriguing research sponsored by the SIHE investigated a more skeptical model of the possibilities of hypnosis as a tool of mind control.

Consider the work of University of Pennsylvania psychologist Martin Orne. An expert on the subject of hypnosis, Orne was doubtful of what he regarded as the naive views that circulated in US intelligence circles during the early 1950s. In his own SIHE-funded work he took the position that hypnosis was less a practice than a "situation"; it was, he felt, less the technique of hypnosis itself than the interpersonal and situational dynamics of the hypnotic experiment that helped overcome the subject's resistance to performing acts contrary to his will. Even in an authentic state of trance, in other words, the subject never fully lost awareness that the situation was an artificial one, and it was precisely that awareness that explained the willingness to relinquish responsibility for his actions. The trance was thus a second-order one, a simulation of a trance, in which the subject was both performer and spectator of his own actions. In a talk he gave at a conference devoted to the phenomenon of hallucination in 1958, Orne made the point by discussing evidence that some hypnotized subjects experienced vivid hallucinations while also retaining an awareness that the imagery was being produced in their own mind. He concluded, "The process of trance-induction itself can be viewed as a gradual increase of potentiality to experience suggested alterations of the environment as subjectively real phenomena."

Orne tested his ideas in a clever variation on the hypnotic experiment. Having concluded that there was little evidence that subjects could be hypnotized into performing acts contrary to their will, he proposed an elaborate deception. A kind of simulation of the interrogation cell, the "magic room," as he called it, employed simple tricks to convince the subject that he had been hypnotized: telling him, for instance, that he was growing warm while using a concealed device to raise the temperature in the room, or suggesting that his cigarettes had a strange taste and then giving him cigarettes containing a bitter ingredient. In that way, the subject's misconceptions concerning the power of hypnosis could be used against him, producing a state in which normal feelings of responsibility were temporarily weakened. The artifice of the magic room thus worked by convincing the subject that the scientist possessed mysterious means of placing him in his power. The "interrogator" used the deception to place the "prisoner" in a "trance" and, once the subject emerged from that state, present him with notes of the interrogation session that seemingly documented his confession. In effect, what was being tested was the experimental situation itself.

Again it is worth underscoring the irony that the founders of the SIHE strove to dispel popular myths of brainwashing that drew on fictions such as *1984* yet wound up funding the investigation of fictive scenarios and simulations. But this is in many ways consistent with the logic of mind control, which originated in the West's darkest imaginings of the enemy yet, as in a hall of mirrors, kept circling back on itself. It is in keeping with this that a SIHE report contains a bibliography of sources that includes, alongside scientific and official reports, memoirs, and biographical accounts, a list of fictional sources relevant to its research: *1984*, Koestler's *Darkness at Noon*, Ayn Rand's *Anthem*, and David Karp's *One*.

Experiments in Sensory Deprivation

As the 1950s dawned, Orwell's dark fable was also on the mind of the Canadian psychologist Donald Hebb. Though Hebb, who early in his life harbored aspirations of becoming a novelist, regarded *1984* as a "dreadful book," he nevertheless turned to it to make a point in the inaugural lecture he delivered in 1950, shortly after his appointment to a position at McGill University. In a thinly veiled critique of behaviorist psychology and its tendency to reduce human motivations to food and sex, he stressed humans' deep-rooted love of mystery, a trait that helped explain their susceptibility to paradox and political propaganda. The point, he suggested, could be illustrated with reference to Orwell's depiction of the role of propaganda, especially radio, in conditioning the citizens of Oceania to accept completely the paradoxical truths that war is peace or freedom is slavery. He soon commenced research in that area.

Hebb was born in Nova Scotia in the same year the Wright brothers pioneered flight, thus making him, he later wrote, an exact contemporary of the age of air travel, a phenomenon whose human effects he would later study. Following the abandonment of his early novelistic ambitions, he turned to psychology, partly inspired by his encounter with the writings of Freud, whom he later described as "the great man of psychology, who taught us how complex mind and motivation can be" (though he hastened to add that Freud's explanations of the sources of that complexity were "absurd"). His graduate work included extensive training in Pavlovian conditioning with a former student of the Russian physiologist. But the formative scientific experiences of his life, as he subsequently related, were a two-year stint at the Montreal Neurological Institute under Wilder Penfield, studying the impact of brain surgery on intelligence, and then several years researching animal behavior, an experience that, he wrote, had given him deep insight into human beings and their social relations. At the end of the 1940s, he took up a permanent position at McGill, where he resumed his former fruitful collaboration with Penfield.* The move occurred just as he was completing his magnum opus, *The Organization of Behavior*, a book that by forging new connections between neurophysiology and psychology had a major impact on the revolution of brain and behavioral science that took place in the 1950s. Its central concepts, among them the notion of cell assembly—according to which repeated firing of associated neurons strengthens the synaptic

---

* The correspondence between Hebb and Penfield was marked by many tokens of their mutual respect. In one exchange Hebb wrote that future historians would recognize that by the year 1950 the MNI was the leader in the field; Penfield responded by thanking Hebb for his contributions to the MNI and adding that "the temporal lobe is the really interesting part of the nervous system at the moment." Hebb to Penfield, January 26, 1951, DH Fonds, 6-0065, 6-0001, McGill University.

connections between them—provided a new basis for theories of learning and subsequently earned it the title of "second most important book in the history of biology" (after Darwin's *On the Origin of Species*).

Under the deceptively modest title "The Psychology of Hocus-Pocus," Hebb's 1950 lecture announced a new era in psychology. For Hebb, the problem with behaviorism was not simply that it was incomplete or wrong; its reductive conception of human motivation was actually dangerous amid the urgent questions that the age of the atom bomb posed for mankind's future. Accordingly, the psychology lab at McGill had begun to undertake research that, building on what he called Pavlov's "epoch-making study of conditioned reflexes," studied the *general psychology* of the dog. Such research showed plainly that food and sex were not the "whole story of a dog's interest in life." That finding held equally for humans. The "oversimplicity" of old ideas went back, he continued, to a false conception of the nervous system, namely, that it was just a set of connections that remained inactive until something from outside stirred it to action. Interestingly, Hebb related the emerging new conception not to EEG but to a slightly earlier invention, the radio. That, he wrote, had laid the basis for a "large-scale revolution in psychology." By picking up and amplifying slight electrical disturbances, radio made it possible for the human ear to hear them; just so, when this method was applied to the scalp of a living person during EEG, similarly slight electrical impulses could be detected in the form of brain waves. Moreover, the signals were broadcast whether the brain was asleep or awake. The living brain was, in fact, never quiet: "It is now quite clear that thought and human behavior does not have to receive some external signal to set it off; and quite clear that man does not remain inert until some primitive need prods him into action, only to slump back into a coma as soon as his needs are satiated."

Hebb went on to stress behaviorism's failure to take account of human complexity, particularly the human love of mystery, which he identified as a fundamental trait, rooted in the properties of the nervous system. Referring to the theory of cell assembly advanced in his book, he argued that the greater efficiency that resulted from synaptic association helped accelerate brain processes, including thought. At the same time, however, that heightened efficiency meant that the brain's capacity was never fully utilized. The resulting lack of full engagement created the ever-present problem of boredom. Anticipating the studies in sensory deprivation that he would soon begin, Hebb noted that the monotony of, for instance, solitary confinement was profoundly disruptive of mental well-being and speculated, "It appears to be an essential part of the system of extracting false confessions in the Communist countries." A prisoner's intense need for stimulation and novelty was a trait that could be easily exploited. Among other things, this helped explain the attractions of mystery and paradox. Citing Orwell's novel, he identified both as essential characteristics of political propaganda and suggested that humans' susceptibility to mystification "may increase as the propaganda becomes more obviously false."

Those insights contained the seeds of Hebb's research into sensory deprivation (SD). Unlike hypnosis, a method with a history dating back to the nineteenth century, SD was very much a brainchild of the 1950s. It quickly became the subject of intense interest following Hebb's first reports on the results of his studies, both within psychology and related fields and in connection with Western investigations into Communist interrogation methods. Such was the fascination surrounding the new area of inquiry that in 1958 a conference was held at Harvard Medical School to take stock of the findings of the previous half decade. It brought together many of the central figures in this book, including Hebb, Grey Walter, Norbert Wiener, Lawrence Kubie, and John Lilly. In his opening remarks, the Harvard neuropsychiatrist Stanley Cobb placed Hebb's studies squarely at the center of the revolution then occurring in brain science, a revolution, he noted elsewhere, whose true dimensions had first come fully into view at the Laurentian Symposium in 1953:*

The last ten years have brought about a complete revision of our understanding of the CNS. One no longer thinks in terms of stimulus

---

* In his obituary for Norbert Wiener in 1964, Cobb pointed to the Laurentian Symposium as a key moment in making this new conception of the brain visible to the larger scientific community.

and response through an integrating switchboard mechanism. The reflexology of Sherrington and Pavlov has been supplanted by a knowledge of reverberating circuits, feedback systems, communications theory, and the central modulation of sensory perception. A new picture of the activity of the brain emerges.

Essential to the new picture of the continuously active brain was research into sensory deprivation (also known as "perceptual isolation"). Cobb credited Hebb's "ingenious idea" of studying perception by means of SD with having opened an immense field of investigation, one with implications for the understanding of consciousness, for medicine (with regard to conditions such as autism), for cybernetic theory, and, not least, for the philosophical question of "reality."

The experimental setup behind Hebb's "ingenious idea" was simple: in specially designed isolation chambers, volunteers lay on couches in conditions intended to produce a reduction in and resultant monotony (or "depatterning") of sensory input. Visual and auditory stimuli were reduced with the help of goggles and white noise. Subjects were encouraged to stay as long as they could. The results, as Hebb wrote in a report for the Rockefeller Foundation (which helped fund his research), demonstrated that much of what was ordinarily taken for granted about the stability and integrity of the self was wrong. "The intactness of one's personality," he wrote, "the ability to think rationally, is dependent on constant feedback of sensations, impressions, experiences from the environment." Prolonged perceptual isolation led to a very rapid disorganization of the personality. Some subjects lasted no more than a few hours before asking to be let out; others were able to remain in the chamber for up to several days. As Robert Morison of the Rockefeller Foundation noted in his diary following a visit to Hebb's lab in Montreal, it was hard to exaggerate the importance of those findings. Not the least of their significance, he observed, was that they called into question whether there was any such thing as an "inner-directed personality," as had been recently posited by the noted sociologist David Riesman.

But Hebb also showed that in the absence of that feedback, the mind did not simply go blank, as many believed would be the case. As he recounted at the Laurentian Symposium in 1953, he and his students observed that subjects emerging from a prolonged stay in a deprivation chamber frequently reported that they had experienced vivid hallucinations, comparable to those observed under the influence of mescal or those reported by Grey Walter in his flicker experiments. He noted that these visionary phenomena seemed to follow a regular course of development from simple to complex. First the visual field changed from a dark color to a light one; next there were reports of dots and lines or simple geometrical patterns; and finally, in a handful of cases, integrated scenes containing "dreamlike distortions." Subjects reported hallucinations of squirrels, spaceships, and other apparitions as well as experiences of "otherness." One felt that there were "two of me"; another that his head had become detached from his body. Such experiences, Hebb went on, were common to certain forms of migraine, as had been described recently by the neurologist Caro Lippmann and earlier by Lewis Carroll (himself, as Lippmann noted, a *migraineur*) in *Alice's Adventures in Wonderland*.

The EEG readings of subjects who underwent SD, Hebb noted in one report, showed an initial lowering of brain-wave activity, such as occurred in sleep. But the onset of hallucinations was accompanied by a shift to rapid frequencies. The effects, as he wrote elsewhere, could be devastating. Depriving a subject for only a few days of the usual sights, sounds, and bodily contact "can shake him right down to the base, disturb his personal identity, so that he is aware of two bodies (one hallucinatory), and cannot say which is his own, or perceives his personal self as vague and ill-defined; something separate from his body, looking down on where it is lying on the bed."* The experience could

---

* According to the British psychiatrist J.A.C. Brown, Hebb's findings confirmed that the boundaries of the self were not given or stable but were acquired as part of a developmental process and, once acquired, those boundaries did not become fixed and enduring; in madness, under the influence of drugs, or in SD, they became fluid and could even dissolve completely. So dependent were people on external stimuli that their withdrawal brought about "devastating results even in a quite short period of time." Brown, *Techniques of Persuasion: From Propaganda to Brainwashing* (London: Penguin Books, 1963), 212–14.

also disturb his capacity for critical judgment and render him highly suggestible. Hebb's original RF report programmatically identified the thinking process with electrical and chemical events in nerve cells: "There is no room here for a mysterious agent that is defined as not physical yet having physical effects. For scientific purposes mind can only be regarded as the activity of the brain, and this should be mystery enough for anyone." Yet in keeping with the argument he had made in his 1950 lecture, he used radio broadcasts about exactly such mysterious agents when he turned to testing the effects of selective exposure to external stimuli on SD subjects.

In that modification of the original experiment, Hebb followed up on conjectures he had made about the possibility of "implanting" ideas in subjects who underwent perceptual isolation. As he had surmised in his 1950 lecture, it proved possible to establish a measurable susceptibility to "propaganda" about ESP, voodoo, and magic presented to subjects under those conditions. Deprived of normal external inputs, their need for stimulation became so great that it overrode their habitual skepticism about such topics. His isolation chamber may thus be seen as an anticipation of Martin Orne's "magic room"—an experimental space whose purpose, it will be recalled, was to test whether the subject could be made to believe (like Winston Smith in *1984*) that he had confessed to an imaginary crime or to otherwise accept a version of reality wholly at odds with his normal perceptions and thought patterns.

Hebb's use of the radio in his propaganda experiment was not incidental, given the Cold War circumstances of his SD research. Orwell's portrayal of the mass media's role in totalitarian societies illustrated how those circumstances shaped contemporary anxieties and scientific research agendas. In a context in which new forms of political indoctrination were portrayed as turning Eastern Bloc publics into assemblages of "captive minds," little more than vessels for state-manufactured versions of reality, it was not surprising that scientific inquiry into the question of "reality"—not only how it was known but also how it might be manipulated and distorted—commanded such interest, nor, it should be added, that the hallucinations often reported by the subjects of SD research became such a major focus of that interest. This was a vital part of the context of the ongoing investigations into the "magic room."*

KUBARK, or "Enhanced Interrogation"

By the time that, at the end of the Dulles era at the CIA, the agency finalized its classified interrogation manual *KUBARK Counterintelligence Interrogation*, the "magic room" had become part of the lexicon of brainwashing. This document surveyed a decade's worth of research conducted under the auspices of MKUltra and flatly concluded that there was no esoteric science of mind control: "There is nothing mysterious about interrogation." For the most part, as Wolff and Hinkle had shown, Communist interrogators relied on a simple if nevertheless highly effective repertoire of methods of stress, isolation, and sleep deprivation.

Yet *KUBARK* also enumerated several unusual methods that American researchers had investigated in their efforts to fathom the seeming riddles of Communist brainwashing. A common feature of these methods was their explicit incorporation of elements of mystery and make-believe. They took as their starting point Hebb's discovery that the radical disruption of familiar patterns of time, space, and sensory perception could combine in a way that "explodes, as it were, the world that is familiar to the subject as well as his image of himself within that world." Hebb's comparison of his findings on SD to the altered states experienced by Alice in Lewis Carroll's stories was echoed in the document's reference to a method called the "Alice in Wonderland" technique. The aim was to disrupt the conditioned reactions of the interrogatee,

---

* A report on one MKUltra subproject contained details about research involving a "special chamber," a "vibrating room," elsewhere referred to as a "magic room." Steven Kinzer, *Poisoner in Chief: Sidney Gottlieb and the CIA Search for Mind Control* (New York: Holt, 2019), 56. The research was overseen by L. Jolyon West, a psychiatrist commissioned by the air force to study brainwashing. West organized the 1958 conference on hallucination at which Orne spoke about pseudo-hallucination.

to replace the familiar with the "weird" by barraging the subject with a cacophony of unrelated nonsensical questions. The final objective was to induce in the subject a state of disorientation, regression, and suggestibility. Regression, the report went on (citing Orne), was also a key objective of the so-called magic room, in which props were used to manipulate the subject into believing that he had been placed in a trance. Equally disruptive were the effects of sensory deprivation. On that topic the document cited the work of John Lilly, the neurophysiologist who in the mid-1950s had cautioned his colleagues against taking 1984 too literally (for more on Lilly, see chapter 7). Isolation and the resulting stress and anxiety, Lilly reported on the basis of his own research into SD, produced effects ranging from hallucination to the perception that inanimate objects were alive. No longer reality bound, the mind turned to fantasy and "projects the contents of his own unconscious outwards." Again, although there was nothing mysterious about those methods, their effects could be greatly enhanced, as Hebb had suggested, by exploiting the subject's susceptibility to mystification.

A notable feature of the resulting state of regression was what the report referred to as the feeling of intense love experienced by some subjects: "The calculated provision of stimuli during interrogation tends to make the regressed subject view the interrogator as a father-figure." By first withholding inputs and then gradually restoring them through a process of selective reward, the interrogator could cultivate in the subject a condition of childlike dependency and gratitude. That finding lent confirmation to Orne's view that the mystery of Communist brainwashing resided less in any specialized methods than in the dynamics of the relationship between prisoner and interrogator.

The "Prisoner's Cinema"

The suggestible, hallucinatory state described in KUBARK has a close kinship with the "prisoner's cinema," the clinical syndrome in which subjects exposed to certain forms of overstimulation (by way of electricity, drugs, or strobe) or, conversely, understimulation (by way of sensory deprivation) experience hallucinations and in some cases memories, however unreliable they might be. Versions of that phenomenon appeared in numerous midcentury accounts, spanning contexts that ranged from Hebb's and Orne's experiments to Edward Hunter's description of brainwashing. Scientific experts pivoted among these various contexts, frequently observing commonalities among them. In 1958, for instance, the psychiatrist and brainwashing expert L. Jolyon West convened a conference on the subject of hallucination that brought together researchers, including Orne, on epilepsy, migraine, and SD. Several of them underscored those commonalities. One speaker described the case of a former POW who underwent SD and then relived, with near-cinematic clarity, his "past horrifying ordeal of being tortured by the Chinese Communists. . . . He saw, heard, believed, and relived the episode." It was on the basis of such accounts that West's colleague Ronald Siegel later coined the term "prisoner's cinema." In describing the syndrome, Siegel referenced Hebb's experiments in sensory deprivation as a crucial point of departure for investigations into this phenomenon.

Though Hebb turned his attention away from such experiments by the mid-1950s, he continued to draw on their lessons for much of his later career.* In the late 1950s, for instance, he evoked them in addressing the question of how the mind knows the world. He referred to what he called the "theory of the imprisoned knower," which had first emerged as a philosophical problem with Descartes and continued to haunt present-day psychology. The imprisoned knower was the thinker who, "encapsulated in his own consciousness," knows the outside world only through unreliable inference. In subsequent writings Hebb illustrated this problem of unreliable inference by invoking the image of a "prisoner shut in a dungeon with five TV screens—the five senses."

---

* Hebb explained hallucinations as "release phenomena": in the absence of sensory input, nerve cells began to fire on their own, and that spontaneous activity could, as a result of the synaptic connections within cell assemblies, sometimes become organized in meaningful patterns, even motion picture–like memories. Hebb, "The Mammal and His Environment," *American Journal of Psychiatry* 111, no. 1 (1955): 826–31; Hebb, "The American Revolution," *American Psychologist* 15, no. 12 (1960): 6.

The fact that the world pictured on the screens is one the prisoner has never been in actual contact with raises the question "How can he know whether the programs are live, or merely taped: how can he be sure there *is* any world such as that that appears on the TV screens?" The solipsism to which the fallacy of the imprisoned knower gives rise could be easily countered, he wrote, by recognizing that the communication system—the nerves—connecting brain to world was not purely mechanical, a mere passive transmitter of signals, but was part of the same substance as the brain itself.

Following Hebb's discussion of sensory deprivation at the Laurentian Symposium in 1953, the Rockefeller Foundation's Robert Morison had invoked the Cold War context of this research in suggesting that the hypnotist or brainwasher substituted his own "choice of sensory patterns for those which should be arriving over the subject's own sensory system." The resulting hallucination was thus triggered by manipulating the subject's relation to the external environment. But as Hebb was arguing by the end of the 1950s, hallucination could occur without manipulation, drugs, or mental illness. Indeed it was entirely normal, arising as the result of isolation or monotony such as that experienced by airline pilots or truck drivers. (He went so far as to say, "Your awareness of yourself, at this moment, is a hallucination that happens to agree with reality.") He connected Penfield's observations regarding his memory patients to his own experiments on isolation, observing that the subjects of such experiments reported seeing vivid scenes "as complex as motion pictures," while also remaining aware of the unreality of the scenes. Hallucination was, he concluded, simply an example of the "improbable tricks" even the healthiest of minds could play on itself.

Such observations were echoed by other authors writing about hallucination and its relation to the enigmas of the human nervous system. Their accounts often included reports that hallucinatory imagery appeared as a movie or slideshow projected onto a screen. Jolyon West's colleague Ronald Siegel collected numerous such accounts, among them the earlier noted case of the former Korean War POW who, in isolation, relived his torture ordeal in cinematic detail. It was likely from Hebb and

West that Siegel derived the notion of the "prisoner's cinema."* He stressed
the importance of research into this phenomenon; it offered, he claimed,
insight not simply into the workings of the brain and into mental illness
but (citing figures such as Socrates, Joan of Arc, and St. Teresa) into the
history-making force of hallucinatory vision. Oliver Sacks, drawing parallels similar to those made by researchers in the 1950s, later underscored
the point that the hallucinations experienced during migraine mirrored
those in sensory deprivation or those experienced by Penfield's epileptic
patients following electrical stimulation of the cortex. Such phenomena
became part of the taxonomy of what one speaker at West's conference,
echoing Hebb's own references to the "Alice in Wonderland syndrome,"
called the most dramatic elements within the "neurological wonderland."

Into the Neurological Wonderland

Having traced this partial itinerary of Cold War mind control, Western
style, let us now return to our starting point. Midcentury researchers, as
we have seen, commonly placed the problem of brainwashing within the
same framework as they did their research into migraine and epilepsy:
it, too, belonged to a wide spectrum of specialized methods for getting
deeply inside a person's head. In some instances, their methods were used
to induce deep alterations in the subject in preparation for the next stage
of the process. Having produced the desired state of suggestibility, they
rendered the subject receptive to the necessary "implant"—whether for
cure or for confession. Once the pattern of her normal conditioned responses had been sufficiently disrupted, the subject was ready to enter a
new perceptual and behavioral state.

How long lasting were the effects of this process? What was the actual nature of the "implant"? What relationship did it have to ordinary

---

*   In 1969, Siegel wrote to Hebb to ask him to contribute to a volume on hallucination,
citing a paper of Hebb's in which he linked the experience of hallucination during SD to
watching a film. Siegel to Hebb, April 5, 1969, Hebb Fonds, 8-0125

perception or existing memory? Was it, in fact, a kind of hallucination or, as Orne suggested, "an experience of suggested alterations of the environment as subjectively real phenomena"? How aware was the subject of the unreality of the new state? No clear answers emerged. Yet figures such as Harold Wolff and Donald Hebb continually circled around the discovery that phenomena ranging from the symptoms of migraine and epilepsy to the effects associated with hypnosis, isolation, and stroboscopic flicker seemed to crack open portals to states of consciousness previously ignored by behaviorists. For Wolff, migraine served as the original point of entry to such states and, more broadly, to the riddles of the relation of brain and mind, riddles including those associated with Communist interrogation as well as the range of possibilities opened up by the SIHE's mission.

Consideration of those possibilities raises questions about the implications—scientific and philosophical, as well as ethical—of opening such portals. Wolff himself was quite willing to take advantage of the opportunities to carry out cutting-edge research presented to him by the Cold War. In doing so, as Marks observes, he and his sponsors operated "far in advance of trends in the behavioral sciences." CIA backing afforded Wolff and others like him both funding and wide latitude for their work. Considered in terms of this book's larger narrative about the history of brain science, the implications were significant: in fields where the behaviorist approach associated with John Watson and his disciple B. F. Skinner had long maintained a kind of stranglehold, the CIA's investment in the research of Hebb, Wolff, Orne, and others, as Marks observed, "helped liberate the behavioral sciences from the world of rats." The debunking of 1984-style mind control was thus also an implicit challenge to the behaviorist paradigm of human science on which that novel's concluding scenes are based. Such challenges contained seeds that later bore fruit in the birth of cognitive neuroscience; it also, as we shall see, found significant offshoots in the counterculture.

Yet the no-holds-barred experimentation associated with that project had a manifestly dark side. Wolff embraced the freedom from ethical constraints necessary to pursue its aims to their logical conclusion with little evident qualm. This is most clearly born out in the support the SIHE provided for the work of Ewen Cameron, the Montreal psychiatrist who combined research into sensory deprivation with an already existing program of investigation into the possibility of treating his patients by first erasing their memories and then implanting new ones. In his indiscriminate use of electricity, drugs, SD, and hypnopedia, Cameron enacted a transgressive will to knowledge that crossed over into nightmare. As became clear in a flood of lawsuits filed against the CIA and the Canadian government in the late 1970s that stretched on for more than a decade, his experiments in extreme "depatterning" left behind a shocking trail of human wreckage: dozens of lives lost to amnesia and other lasting effects of his determination to wipe his patients' brains clean. In committing those serial lapses in professional ethics Cameron was, of course, far from alone. Long after the formulation of the Nuremberg Code at the end of World War II, American medicine continued to be plagued by blind spots with respect to patients' rights. Those blind spots, as we have seen, were central to the critique articulated by the antipsychiatry movement that first emerged in the 1950s and found expression in patient narratives such as that authored by Barbara O'Brien, the author of *Operators and Things* (see Clinical Tale 2). O'Brien's narrative is echoed in many details by that of Velma "Val" Orlikow, a patient of Cameron whose husband successfully sued the CIA and Canadian government in the 1980s. Orlikow's narrative is illustrative of the disconcerting tendency noted by the authors of *KUBARK*, namely, the fact that she, like a number of Cameron's other largely female patients, reportedly worshipped and adored him even while suffering deeply and in some cases irreparably at his hands. Orlikow reported the terrifying results of being treated in Cameron's "sleep room" in the following terms: "You become very small. You're going to fall off the step, and God, you're going down into hell because it's so far, and you are so little. Like Alice, where is the pill that makes you big."*

---

* A chronology of New York Hospital's involvement in MKUltra dating from the 1970s bears partial witness to that tortured history. Though it documents the broad outlines of the program Wolff oversaw, the account is full of gaps and silences. On the question of whether unethical experiments were carried out there as part of the program, the answer is a somewhat hedged "no." "A Brief Chronological Outline of Activities Carried Out at Cornell University in Response to a Request Made by Mr. Allen Dulles in 1953–54." Wolff Papers.

Conclusion

In his willingness to go to extremes and to fund those similarly willing, Harold Wolff was in many ways representative of his scientific caste. The combination of political pressure to produce results and the professional opportunities that came with such pressure created powerful incentives that acted to weaken ethical restraints. Not all scientists, however, embraced such opportunities unhesitatingly. Apparent exceptions include two figures closely associated with the history of research on sensory deprivation, Donald Hebb and John Lilly. Hebb's case has attracted much attention. There is little question that he initially worked with the Canadian (and indirectly US) intelligence services or that his work spawned numerous imitators. Yet he had grave misgivings about the secrecy requirements surrounding that work. Originally prevented by the agencies that helped fund his sensory deprivation research from alluding to the Cold War circumstances surrounding its origins, Hebb remained anxious to dispel the secrecy, and he seized the opportunity to do so. In the opening address to the symposium on SD held at Harvard in 1958, he stated, "The work we have done at McGill University began, actually, with the problem of brainwashing." Though he inhabited the same Montreal scientific milieu as did Ewen Cameron, he regarded Cameron with distaste and his work on "psychic driving" as grotesque. The degree to which Hebb himself managed to resist the pressure to violate experimental ethics has been disputed, though he maintained that his studies were conducted under conditions in which subjects participated on a strictly voluntary basis and would be released from the chamber anytime they wanted.

The response of John Lilly, who pursued Hebb's research to its farthest extreme, to these dilemmas was to restrict his experiments in sensory deprivation largely to himself. Though he proved in many ways to be the maddest of this cohort of midcentury scientists, Lilly was, at least in that regard, a figure of some restraint. In urging his colleagues at the GAP symposium on forced indoctrination that was held in 1956 to distance their investigations from the scenario depicted in Orwell's *1984*, he sought to undo the spell that such scenarios had cast over the midcentury imagination. Paradoxically, as we shall see in the final chapter, that eventually led him to even more fantastical scenarios, in which Orwell's dystopia continued to serve as a kind of limit case of both the Cold War and the countercultural imagination. In an unpublished text from around 1960 titled "Human Brain Research and Ethics," Lilly conjured up dark possibilities, asking "Should a Manhattan Project be set up on mind control by electrical control of the brain?" He then warned of the need for appropriate preventive measures to avoid a future in which the visions of *Brave New World* and *1984* might be realized. I will return to Lilly later, but first will turn to a consideration of how the investigations of Cold War–era brain scientists found their way into the broader cultural realm of midcentury America.

Even as Harold Wolff and Lawrence Hinkle sought to demystify Communist interrogation practices, evidence continued to be brought forth of their fundamentally diabolical nature. The year 1956 marked in some ways the high point of the war of conflicting representations over Red mind control, with the near-simultaneous release of the Wolff-Hinkle report and Edward Hunter's updated account of brainwashing, heavily reliant on claims about the role of Pavlovian science. Toward the end of that same year, a new and especially lurid account was introduced into the public record. Its source was a young Hungarian dissident, Lajos Ruff, who'd left his country after the failed 1956 uprising and made his way to Vienna. There he encountered and was interviewed by William Rusher, an up-and-coming member of the conservative movement who was employed as associate counsel for a Senate subcommittee on Internal Security. Arrangements were hastily made to take Ruff to the United States to deliver testimony on the phenomenon of "so-called 'brainwashing.'"

Under questioning conducted by Rusher, Ruff laid out for the assembled members of the subcommittee a tale of betrayal, arrest, and sadistic treatment at the hands of his captors. In August 1953, at the age of twenty-three, he had been detained and charged with plotting against the regime and distributing anti-Communist leaflets. Initially he had been subjected to common methods of torture, such as being forced to stand in cold water for hours at a time. When that had failed to produce the desired results, he had been taken to a "psychological investigation room" where special methods had been used on him over a period of approximately six weeks. As though borrowing from the plot of a B horror film, Ruff supplied copious fantastical details. Among other features, the so-called bewitched room was equipped with lamps with perforated shades that revolved continuously on the ceiling, casting spiral patterns and creating other weird optical effects. "Special films" were also screened on the walls of the room, and a psychiatrist who had reputedly also treated Cardinal Mindszenty in the same room injected him with scopolamine and mescaline. Ruff reported that all prisoners who underwent this process eventually became "schizophrenic" but that he himself had feigned insanity and had been transferred to an asylum. Eventually, following a trial, he had been convicted and sent to a prison, where, he related, he had encountered Mindszenty, the figure who had become emblematic of the horrors of Red mind control. On November 1, 1956, amid the Hungarian Uprising, he was released, and he quickly made his way to the West. Six weeks later, he delivered his testimony to a committee made up of leading members of Congress.

Ruff's sensational narrative was expanded into a book, *The Brain-Washing Machine* (1958), which he had written, so the dust jacket claimed, in a room at the Library of Congress.* Originally published in Paris, the book was translated into ten languages and brought its author worldwide attention. Ruff's account was in a sense an updated version of his countryman Arthur Koestler's by now classic *Darkness at Noon*, the difference being that Koestler's novel, although based on historical

---

* US State Department documents, which refer to him as "Ludwig Ruff (also known as Lajos Ruff)," show that he stayed only briefly in the United States before returning to Vienna. The circumstances of his return were complicated by the hasty way in which US officials had spirited him out of Vienna to the US capital. Department of State telegram, March 7, 1957, National Archives II.

events (the Soviet show trials of the 1930s), did not hide its status as fiction. The clearly embellished nature of Ruff's account, however, was passed off as a factual narrative retold with complete fidelity by a man whose exceptional powers of endurance had, according to the book's dust jacket, enabled him "to retain a coherent memory of the methods employed." Ruff's account was a self-conscious hybrid of totalitarianism discourse—it supplied proof of Hannah Arendt's claims regarding "Bolshevism's laboratory techniques for producing confessions"—and B-movie plot details and special effects, details that led one Hungarian reviewer to dub it a piece of "émigré pulp fiction." The CIA's in-house publication *Studies in Intelligence* suggested that in its indiscriminate piling up of details, Ruff's account explained "too much" and that it was best understood as a "mescaline dream."* Yet in other parts of the US government (and elsewhere, as we shall see), many people were prepared to take it seriously.

In the book's key chapter ("Destruction of the Soul"), Ruff described the "magic room" in the following terms: it was "the most frightening workshop of Soviet mental destruction, a psychological atomic reactor. . . . Just as the physicists of the twentieth century penetrated into the smallest particles of matter and had learned the most guarded secrets of nature, so the magic room penetrated the infinitely finer construction of the human soul, but only to destroy the soul's hidden forces after discovering them." All those who entered the chamber, he said, were reduced to a state of helpless self-denunciation and ultimately to schizophrenia.

As with the hypnotist Martin Orne's "magic room" or the Alice in Wonderland technique described in *KUBARK* the chamber was designed to render the subject completely suggestible. Among its many special features were its irregular contours (a wall with an oval shape

---

* In writing it off as fiction, however, the review explained too little. As we have seen, Cold War fiction played a vital role in providing pseudo-empirical validation of phenomena, such as the so-called mind control gap, which remained largely a matter of speculation. "Book Review: *The Brainwashing Machine*," *Studies in Intelligence*, Spring 1960.

and a bed sloped in such a way as to render sleep difficult) and unusual lighting fixtures. Two continuously rotating lamps, whose shades were perforated with small holes, "projected fantastic lights and colors into the room." The stroboscopic effect produced progressively more severe visual disorientation. Meanwhile, personnel further played with his sense of reality by, for instance, awakening him from his fitful sleep and accusing him of trying to commit suicide, then "finding" props such as a silk scarf in his bedding to confirm their allegations. Next came regularly administered drugs (scopolamine and mescaline) and interminable conversations with the psychiatrist who oversaw the proceedings, Laszlo Nemeth. Following one such conversation, which endlessly rehearsed the details of Ruff's life, Nemeth departed with a friendly warning to "avoid the silver beam. It has an unusual effect on the brain." The beam flooded the room from an aperture in one of the walls, and Ruff spent hours fruitlessly trying to avoid it before finally entering a kind of trance state in which, he stated, "I even desired the silver beam."

A further stage in the process was reached when he once again awoke from fitful dreams to discover that moving pictures were playing on a screen on the wall. The images, mostly erotic in content, continued for hours, varying in tempo, sometimes accelerated, sometimes slowed down. As Nemeth explained to him at one point, the films showed the way a schizophrenic patient sees the world. On one occasion, after watching a pornographic film, Ruff awoke to discover the actress in the film lying in bed next to him. She continued the scene with him. Following their lovemaking, Ruff wrote, "It was impossible to say what was real and what was the beginning of hallucination."

Inevitably, he broke down, surrendering the secrets he had managed to guard throughout his earlier physical torture. Mere confession, however, was not enough; in order for the story to stand up in the courtroom, he had to be made to believe his confession fully. As part of the lengthy preparation, he was subjected to the final modification of the clinical syndrome we encountered earlier as the "prisoner's cinema." New films were screened in which he starred as the hero of the Hungarian Republic, becoming both an actor in and a spectator of false memories. Living in the wholly imaginary world of the films projected onto his wall, he finally came to believe every detail of his confession. The "ray-treatment" of the silver beam also played its part, whether through paralyzing the will or "damaging the brain." Having arrived at that point, he was ready for the show trial.

Yet Ruff was ultimately saved when he accidentally broke the lamp and became suddenly aware of the awful peril confronting him. In a fit of simulated madness he proceeded to smash all the other props in the room. The purpose of the diabolical chamber, he realized in a flash of insight that echoed the findings of Orne's "magic room" experiments, was to alleviate its occupants of "the burden of free choice." His own personal revolt against his captors, he concluded, was later acted out on a larger scale in the Hungarian Uprising of 1956, sparked by people who had undergone "the longest treatment in the magic room of Communist propaganda and ideology behind the Iron Curtain."

The fact that Ruff's account had an ambiguous relationship to historical reality, yet might nevertheless possess truth value, was the conclusion reached by more than one of its readers. An interesting example may be found within the response it generated in parts of the French avant-garde. As noted above, the book was originally published in Paris, where it appeared in the same year as the first published (albeit quickly censored) accounts of the torture practiced by French troops during the Algerian War. In June 1958, the inaugural issue of *Internationale Situationniste*, the newly founded publication of the Situationist movement, included an article titled "The Struggle for the Control of the New Techniques of Conditioning" that treated Ruff's account with a combination of skepticism and seriousness. According to the author of the piece (probably Guy Debord, later the author of the book *The Society of the Spectacle*), the methods Ruff described belonged to a spectrum of new possibilities ranging from subliminal advertising to sensory deprivation, all of which sought ways of influencing human behavior. The piece began by citing the Russian propaganda expert Serge Chakotin, the author of *The Rape of the Masses by Political Propaganda* (1939), an influential work that adapted the ideas of Chakotin's former teacher Pavlov about reflex conditioning to the analysis of modern propaganda. Borrowing directly from Chakotin, Debord's first line announced, "It is now possible for human reactions to be triggered in a predetermined direction." It then proceeded to cite several of the scientific experiments conducted in the field.

Of particular interest was the author's reference to the McGill experiments on sensory deprivation and the mental disturbances that had resulted. The author then went on (after duly registering some skepticism) to cite many details of Ruff's description of the "magic room." For that author, it is clear, it scarcely mattered whether Ruff's account was factually correct or completely fictitious; what mattered was its predictive or diagnostic value. In a manner that accorded perfectly with the logic of the Cold War, it could be treated skeptically yet still taken deadly seriously as a warning, a predictor of possibilities or scenarios that might yet become all too real (not unlike those described by Aldous Huxley in *Brave New World Revisited*, published the same year). In a final surprising twist, the text concluded by recognizing the emancipatory potential of mind control techniques: "It should be understood," Debord concluded, "that we plan to dive headlong into *the race between free artists and the police to experiment with and develop the use of the new techniques of conditioning.*"*

---

* In another version of the syndrome of the "prisoner's cinema" that was experienced by Lajos Ruff, Debord's *The Society of the Spectacle* (New York: Zone Books, 1994 [1967]) described modern society as a spectacle whose "spectators" were "imprisoned in a flat universe bounded on all sides by the spectacle's *screen*" (152–53).

PART III

PAUL LINEBARGER/ CORDWAINER SMITH

The image of the electronic brain that first appeared in the writings of cyberneticians such as Norbert Wiener exercised a powerful hold over the midcentury imagination. In early 1950, it was represented in a strangely anthropomorphic form on the cover of *Time* magazine, accompanied by an article on advances in computing. Soon thereafter, it found its way into Edward Hunter's first account of brainwashing (1951). Figures as diverse as Hannah Arendt, Barbara O'Brien (the author of *Operators and Things*), and the media theorist Marshall McLuhan also invoked this strange apparition. At the Laurentian Symposium, the psychologist Donald Hebb mused about the possibility of training an electronic brain to introspect. As we turn in this section to look at the way new understandings and representations of the brain entered the wider culture of the 1950s, we encounter versions of this image in many places. It is not surprising that it can be found in midcentury science fiction or in new forms of healing such as Dianetics, the hugely successful movement that had itself emerged from science fiction and that borrowed liberally from the contemporary sciences of brain and mind.

What is perhaps remarkable is that all three of these quintessentially midcentury types—psywar specialists, purveyors and disciples of new healing doctrines, and science fiction authors—could coexist in the same person. Such was the case with Paul Linebarger or, as he was known to science fiction readers, Cordwainer Smith (among other noms de plume). Linebarger is one of the most chameleon-like presences in this entire story, a figure who moved back and forth among academia (he taught Asian Studies and seminars on psychological warfare at Johns Hopkins University), the highest levels of the US intelligence community, the world of Dianetics (later known as Scientology), and the same pulp sci-fi milieu out of which had emerged his friend Hubbard (whose first contribution to the genre Linebarger had published back in the 1930s, when both were students at George Washington University).

Linebarger's main claim to importance rested on his book *Psychological Warfare* (1948; revised 1954), long considered the standard text in the field. Though the majority of the book was devoted to discussion of propaganda operations, Linebarger also covered a range of other practices that fell more properly under the rubric of "brain warfare." In the revised edition, for instance, he devoted attention to the phenomenon of brainwashing. He credited his friend Edward Hunter with having first revealed the practice to the world, although one scholar has suggested that Linebarger himself may have coined the term. Aware of the skepticism surrounding Hunter's account, Linebarger took pains to say that he had known Hunter for twenty years and vouched for him as a man with "sober respect for fact." He described the method as a "frontal attack on all levels of the personality" in which endless interrogation culminated in a "nervous breakdown," followed by a Pavlovian process of rebuilding the prisoner's psyche.*

In a speech given at the US Naval War College in 1950, Linebarger expanded the field of psychological warfare to encompass what he called "psychiatric warfare," in which conventional war was replaced

---

* In 1956, Hunter wrote to Linebarger to thank him for the glowing review he had given his new book on brainwashing in a State Department report for overseas librarians. Hunter to Linebarger, September 15, 1956, Linebarger Papers, Hoover Institution Library and Archives, Stanford, CA.

by an organized campaign of attack directed at the enemy's personality. He illustrated it with reference to the case of the Hungarian cardinal Mindszenty, speculating that it likely involved some combination of psychopharmacology and hypnosis. Something very similar, he argued, was going on in the forms of brainwashing used by the Chinese. Linebarger was stationed in Korea for part of the war and incorporated his observations made there into the revised edition of his book that appeared in 1954.

A participant in one of Linebarger's psychological warfare seminars later credited him with adopting a kind of psychiatric approach to the process of understanding "black propaganda operations," that is to say, operations designed to look like acts of the Communist enemy and thus to sow confusion in its ranks. In the evening seminar meetings held in his Virginia home, at which students were encouraged to dream up schemes for attributing the United States' charge of Chinese brainwashing to neutral countries such as India to lend them more credibility, Linebarger used a kind of "group therapy" to show his students that tricking others into believing, for instance, that black was white was a natural ability. In what the author referred to as the "class confessional," Linebarger challenged his students to remember incidents from their childhood illustrating that ability: "Try to remember how old you were when you first tricked one of them" (i.e., a family member). Each seminar meeting opened with twenty minutes of such confessions, in which participants retrieved long-buried, and perhaps shameful, parts of their childhood. Linebarger, his student wrote admiringly, was more inventive in applying behavioral science to psychological warfare than anyone else he had ever met. Yet such exercises were only one facet of the complex role that memory and its enigmas played in Linebarger's life.

The reputation that his expertise in the field brought him led to Linebarger's appointment to the Psychological Strategy Board, a committee made up of representatives of the State Department, the Department of Defense, and the CIA to oversee US intelligence and propaganda operations (later reorganized under the title Operations Coordinating Board, or OCB). There he helped advise the Eisenhower administration on psywar campaigns both at home and abroad. It was from that perch that he likely played a role in midwifing one of the most lurid contributions to the literature of the Cold War, "Brainwashing: The Communist Experiment with Mankind," along with an accompanying document analyzing a text on "psycho-politics" allegedly written by the head of Stalin's secret police, Lavrenty Beria (see chapter 3). Rumors long swirled around the authorship of the texts, with some people believing that L. Ron Hubbard had had a hand in them. In the early 1960s, a memo circulated among members of a congressional subcommittee in which confirmation was sought of Edward Hunter's belief that the text had been written by Hubbard. But circumstantial evidence, not least his membership in the body most directly concerned (the OCB), points to Linebarger's role in crafting the documents or at least facilitating their circulation at the highest levels. The "Beria" text in particular bears many of the hallmarks of exactly the kind of "black propaganda" operation that so fascinated Linebarger. (The confusion about the origin and authenticity of the documents attests to the fact that the different branches of the US national security apparatus were not always fully informed of one another's actions.)

A key passage in the first document, whose depiction of the brainwashing process is highly novelistic, described the Pavlovian techniques being used by the Soviets on their populace. An endless stream of manifestos, newspapers, speeches, plays, radio broadcasts, and movies transmitted suggestions intended to condition their reflexes: "The suggestions and words become strong engrams, work their way into the brain cells, become part of the subject's thinking and therefore a power in the hand of the evil controlling the propaganda."

The use of the term *engram*, used by some midcentury psychologists in reference to a postulated neural memory trace, is one clue pointing to the influence of Dianetics. As Linebarger explained in his unpublished text "Ethical Dianetics: The Lay Psychotherapy of Mutual Aid" (c. 1952), the term *engram* was central to the science of the mind and of healing invented by L. Ron Hubbard at the beginning of that decade and eventually rechristened as Scientology. Dianetics, one of the many mutant offspring of cybernetics, was first introduced to the world in 1949 in an announcement in *Astounding Science Fiction*, a publication in whose pages Hubbard had been publishing his brand of pulp literature for several years. The announcement, which was penned by the magazine's editor, John Campbell, presented Dianetics, whose name played on Wiener's coinage "cybernetics," as a revolutionary new science of the human mind.* Campbell had studied under Wiener at MIT in the 1930s, and he was fascinated by the analogy his former teacher had drawn between computing machines and human brains. It was at Campbell's urging that Hubbard incorporated the analogy into the first publications on Dianetics. As the philosopher S. I. Hayakawa wrote in a review in 1951, Hubbard's subsequent success reflected the powerful contemporary allure of mechanical analogies at a time when even the most casual newspaper reader was beginning to "hear of 'brain waves' and 'electroencephalograms.'" (Hubbard's error, according to Hayakawa, lay in ignoring the fact that they were simply analogies.)

When Hubbard contacted him in 1949 about his new discovery, Campbell responded enthusiastically and immediately volunteered to undergo "auditing." Campbell hoped that Hubbard's new technique would help unblock memories of his childhood, but Hubbard's usual methods, hypnosis and drugs, failed to help. Success came, however, when they switched to a new approach involving a device whose elements consisted of a lit candle and four mirrors arranged in pyramid form on a phonograph turntable. Once the turntable was switched on, the device produced a light flashing three hundred times per minute. The effect on Campbell was dramatic: it unleashed a flood of memories that left him shaken and overwhelmed. The device produced effects very similar to those produced by Grey Walter's experiments with stroboscopic flicker. This was no accident. In an unsent letter Campbell wrote in 1953 to Walter (whose book *The Living Brain*, published that

---

* Wiener considered filing suit against Hubbard for his misappropriation of "cybernetics." His papers include a letter from Hubbard apologizing for any offense taken at efforts made by figures in Hubbard's circle to link Dianetics with cybernetics. Hubbard to Wiener, July 26, 1950, Norbert Wiener Papers, MIT, 8, 121.

year, he had read closely), he recalled the experience. Without mentioning Hubbard, whom Walter had disparaged in his book, Campbell discussed his inability to remember his childhood and the failed experiment with hypnosis: "I was about as suggestible as one of your EEG machines."* He then described the startling results obtained by means of the mirror device, which, he surmised, opened "channels of memory pathways to past experiences I hadn't been able to reach." He expressed his belief that the flickering light matched his alpha rhythms and "compared its effects to electroshock therapy or the use of drugs to induce seizures in psychiatric patients." As we shall see in chapter 7, the device anticipated the Dream Machine, which, likewise inspired by the writings of Grey Walter, was invented at the beginning of the 1960s by friends of William S. Burroughs, another figure deeply intrigued by Scientology.

Hubbard's book *Dianetics: The Modern Science of Mental Health* appeared the following year and became an immediate best seller. It fell on fertile ground at a moment marked by widely shared anxieties about the methods of midcentury psychiatry and the onset of the Cold War. So successful was it that Grey Walter, in *The Living Brain*, warned his readers about the hold that gurus such as Hubbard were acquiring over their followers. He saw in Dianetics a phenomenon that had succeeded in reducing all previous forms of healing (Freud, Jung, Pavlov, behaviorism, and so on) to "the lowest common denominator." Most vexing from Walter's point of view, Hubbard had created a doctrine that naively asserted the claims of mind over brain.

Hubbard's approach rested on what was in essence a highly simplified form of psychoanalytic treatment that promised to liberate followers from all their ailments, both mental and physical, the sources of which were ultimately traceable to buried memories. The so-called engrams were believed by Hubbard to exercise control over individuals

---

* Hubbard and Campbell seem to have experimented with using EEG during auditing before replacing it with the so-called E-meter. See Alec Nevala-Lee, *Astounding: John W. Campbell, Isaac Asimov, Robert A. Heinlein, L. Ron Hubbard, and the Golden Age of Science Fiction* (New York: Dey Street Books, 2018), 329.

like posthypnotic commands or tape recordings that looped endlessly through the mind. As he put it, the individual is "handled like a marionette by his engrams." The reexperiencing of the trauma encoded in the "latent memory picture" of the engram would, he claimed, result in the erasure of the individual's Reactive Mind (a version of the Freudian unconscious). Upon completion of the process, the individual would be pronounced "clear."

Yet liberation from the power of so-called demon circuits was only one part of the Dianetics program. Even as Hubbard cobbled together elements of Freud, Pavlov, and experimental psychology to fashion his new science, Dianetics began to acquire the trappings of a religion. In part that was to avoid trouble with the IRS, but in part it flowed naturally from the messianic aspects of Hubbard's personality. As Hubbard wove a vast amnesia narrative dating back to the origins of humanity, the memory project at its heart took on cosmic dimensions. The central actors within this narrative were his professional nemeses the psychiatrists. From the very beginning, as he alleged in one article, representatives of that profession had been harassing him, joining forces with Communist agents in their efforts to take him down. Eventually the mass media became part of this paranoid scenario as well; projecting vast, malign powers onto the psychiatric profession, Hubbard accused its members of forcing victims to "sit in front of movie screens for 36 days of programming."

Virtually from the moment it was first introduced to the public, Dianetics was a highly schismatic movement. Many of Hubbard's most enthusiastic early supporters were either expelled or left on their own to form splinter groups. One such dissident was Paul Linebarger. At some point in the early 1950s, writing under the pseudonym Carmichael Smith, he penned an unpublished contribution to the literature of Dianetics, "Ethical Dianetics: The Lay Psychotherapy of Mutual Aid." This text made plain that his interest in the place of psychology in medicine was of a piece with his interest in the applications of psychology in politics or warfare. Here, too, the personal dimension of his interest became manifest: "I write this book," he announced in the opening pages, "as one layman to another—indeed, as one sufferer to another."

Without specifying the nature of his suffering, Linebarger/Smith made clear that he had found much relief and inspiration in Hubbard's new science.* At the same time, he felt that certain aspects of it were not above reproach. He suggested to his reader that Hubbard, whom he had known since their student days, was like the seventeenth-century physician Franz Mesmer in that both encumbered their discovery of important new dimensions of human behavior in overly elaborate systems of scientific speculation. More specifically, he continued, "Hubbard sets up an imaginary parallel between the human mind and the so-called electronic brains, which are one of the most provocative discoveries of American engineering." In particular, he borrowed the notion of a "memory bank," in which engrams were deposited and from which they continued to exercise an influence over the life of the individual. The engrams were reactivated by key words relating to the original traumatic situation. The process of reliving the original occurrences unfolded in stages: the scene was gradually filled in, first in black and white, then in color, before progressing to a "moving pantomime" complete with sound and smell. (In an aside, Linebarger described this motion picture–like "moving pantomime" as being far richer than the motion pictures churned out by Hollywood.) Though quite ready to credit Hubbard with important advances, he was also careful to underscore areas where he disagreed with his teachings. What most disturbed him was the increasingly doctrinaire and authoritarian nature of the demands Hubbard placed on his followers. Distancing himself from Hubbard's antipsychiatric diatribes, he wrote that Ethical Dianetics, by contrast, was "not a closed cult."

If in that text Linebarger took pains to present himself as offering a sensible alternative to Hubbard's more far-fetched doctrines, in the science fiction stories and novellas that he penned under the name Cordwainer Smith, he gave free vent to an imagination quite

---

* A recent article has described Linebarger as suffering from a range of physical ailments, including near blindness as well as chronic metabolic disorders, and further claimed that he had had suicidal impulses. Ashley Stimpson and Jeffrey Irtenkauf, "Throngs of Himself," *Johns Hopkins Magazine*, Fall 2018.

as fantastical as Hubbard's. Indeed, some of his early ventures in the genre were rejected by John Campbell, the editor of *Astounding Science Fiction*, for being too extreme even by the standards of a genre that trafficked in the outlandish. Such was the case with his story "Scanners Live in Vain," later heralded as one of his finest efforts, in which space travelers undergo a kind of surgical version of sensory deprivation that removes all of their perceptual organs except those of vision in an effort to help them survive the rigors of deep-space travel.*

The strange story of Paul Linebarger's life would not be complete without mention of the rumors, long standing in science fiction circles, that he was the real person behind the pseudonym "Kirk Allen," a figure who appears in one of the clinical tales presented in the psychoanalyst Robert Lindner's best-selling book *The Fifty-Minute Hour* (1954), a collection of case histories of several of Lindner's patients. According to Lindner, Kirk Allen was a scientist who held both a high-level national security clearance and a deep conviction that he inhabited an alternate reality. As a child he had discovered a science fiction series with whose hero he shared the same name. "He identified with the character and, believing the stories were his own biography, began inventing new adventures that he believed he was 'remembering.'"

"Remembering," as this case history reminds us, was both a key midcentury preoccupation and a human faculty whose stability and integrity were being challenged on numerous fronts. The sources of that challenge were many. In an essay on cybernetics, Hannah Arendt returned to the talismanic status of memory, observing that the erasure of memory in a computer (or "electronic brain") was normal but that in a person it constituted a form of brainwashing. Linebarger might well have concurred on that point. Yet as a product of the national security state and a key theoretician of the dark arts so essential to that state's operations, he himself played a central role in devising techniques of

---

* Linebarger's science fiction manifested a fascination with mind and brain that his editors saw as problematic given, as one explained, the fact that the genre was drowning in "psi" (a preoccupation with telepathy and parapsychology). (Horace Gold to Paul Linebarger, November 14, 1955, Linebarger Papers).

psychological warfare whose consequences for public knowledge and memory would long be felt in American society. As in a hall of mirrors, the confusion such operations were designed to sow among the enemy produced similar confusion at home, leaving many Americans, to paraphrase Thomas Powers, to wonder how they knew that what they knew was true.

CHAPTER 5

Manchurian Candidates

The Manchurian Candidate (1962).

Did the human brain change in the 1950s? The question is not as odd as it may at first appear. In the foreword that Aldous Huxley added in 1946 to the second edition of *Brave New World* (first published in 1932), he boldly foretold such change. He predicted that "the most important Manhattan Projects of the future will be vast government-sponsored inquiries into what the politicians and the participating scientists will call 'the problem of happiness'—in other words, the problem of making people love their servitude." Solving the problem, he said, would necessitate a "deep, personal revolution in human minds and bodies." Twelve years later, in the updated version of the novel, *Brave New World Revisited*, he wrote that breakthroughs in psychopharmacology and other fields had brought considerable progress in advancing that revolution.

During the 1950s, many commentators agreed with Huxley that such a revolution was very far along. They worried that postwar consumer society and the forces it had unleashed were bringing about deep changes in brain and behavior. That was especially true in relation to the young, whose comic book–reading and TV-viewing habits were closely tracked by specialists in many fields. Among them was the neurophysiologist Grey Walter. He felt that contemporary social development, which had once seemed to point to a new era harnessing the full creativity of the brain, now appeared to be reverting to the more primitive level of the spinal cord, favoring simple reflex coordination at the expense of independent or original thought: "A passive solitary child gazing at the screen of a television receiver amuses only itself—the need to gaze does not promote or evoke habits of creativeness or generosity." In his capacity as brain specialist, Walter served in many public roles, among them as member of a British committee set up to study the relation between the mass media and juvenile delinquency.

Walter was representative of a new species of expert that emerged at that time, claiming public authority based on knowledge of the brain and its workings. Another such figure was Fredric Wertham, who became widely known in the 1950s for his strident warnings about the ill effects of popular culture on the juvenile mind. In books, Senate testimony, and other venues, Wertham repeatedly stressed the dangers that comic books and TV posed to "young nervous systems." His claim to expertise on the subject was based, as he stated in congressional testimony, on his background as a psychiatrist and brain researcher dating to the 1930s: "Some part of my research at that time was on paresis and brain syphilis. It came in good stead when I came to study comic books." Though he had given up active research in the area, his personal papers indicate that he continued to follow developments in the field. They include clippings of the work of Penfield and Walter, as well as extensive files on Pavlov following the revival of interest in his work in the early 1950s.* Though the nexus between the mass media and juvenile behavior remained an elusive one, Wertham's claims illustrate the growing recourse to brain science in relation to this as well as other hotly debated questions of the day.

The revolution predicted by Huxley may have come closest to realization through the psychopharmacological breakthroughs of this era. For millions of individuals afflicted with disorders such as schizophrenia and depression, the advent in the 1950s of a class of drugs that acted directly on the brain (chlorpromazine, Miltown, and so forth) was a life-changing development. At the same time experimentation with the new class of hallucinogens for which Huxley adopted the term *psychedelics* also affected brain chemistry, albeit in a less easily measurable fashion. The combined effect of this new cornucopia of drugs, together with a host of techniques ranging from neurosurgery and electroencephalography to stroboscopic flicker and sensory deprivation, was to turn the brain into a site for wide-ranging forms of experimentation.

The new picture of brain activity that emerged from those investigations soon spilled into the realm of popular representation. News of findings in midcentury neurosurgical theaters and EEG labs gave the brain both a new public visibility and a new form. In films of the 1950s, brain science is often treated as mad science. Consider, for example, two films of that era, *Donovan's Brain* (1953) and *The Brain* (1962). Both were adaptations of Curt Siodmak's pulp novel *Donovan's Brain* (1942), a tale containing what is likely the first literary representation of EEG. The German émigré Siodmak turned his countryman Hans Berger's invention into the basis of a lurid mind-control fantasy that 1950s filmmakers found irresistible. In it, the brain of the dead industrialist Donovan is removed from his skull and kept alive in a vat, its signals studied closely by the ambitious scientist Patrick Cory: "The electroencephalograms marked their slow, trembling curves on the paper strip which continuously

---

* Included are what appear to be notes for a manuscript on Pavlov's legacy. Fredric Wertham Papers, Library of Congress. It was only natural, as we shall see, that Yen Lo, the Pavlov specialist in *The Manchurian Candidate*, invoked Wertham's name in support of the notion of mind control.

flowed from the wave-recording machine." Soon, however, the scientist goes from studying the brain waves to being controlled by them. The film adaptation's representation of the brain is worth noting: in this as well as in a host of other science fiction and horror movies, the brain, seen by the ancients as a humble, formless mass, acquired a new, more charismatic image. Freed from the confines of the skull, it is revealed as a singular presence: well formed, highly articulated, pulsing away in its tank like a half-submerged creature in a murky lagoon, lit up by an accompanying glow, the by-product of the signals emanating from within that are read by the ever-present EEG.

Popular representations such as these illustrate how breakthroughs in brain science entered mass culture. Some of these developments also became an essential part of the 1960s counterculture. In popular culture as well as in journalistic and sociological accounts and experimental film and fiction, the implications of the new brain paradigm that emerged in the first decade of the Cold War were explored. In those texts and films the brain was treated as an object of immense fascination and ambivalence—an organ possessed of extraordinary capacities but also vulnerable to devastating injury, manipulation, and deception. Damaged brains, hallucinating brains, amnesiac brains, but also superbrains, brains blessed with astonishing powers of recall or telepathic abilities: all were represented repeatedly. Running through such accounts was the ever-reiterated theme of brainwashing, a phenomenon that was as often as not now associated with Western society. In the landmark film *The Manchurian Candidate* (1962), the specter of Red mind control was treated as simply one manifestation of a far more widespread condition common to both Communist and capitalist societies.

The "Dark Arts": Variations on a Theme

Postwar America's emergence as a new phenomenon in world history—the world's first true automated mass consumer society—was a dramatic development that elicited much commentary. From across a wide spectrum, a diverse crowd of authorities stepped forth to reflect on that occurrence and to speculate about its implications for human behavior, identity, and morality. Let us examine some of those accounts, taking the year 1951 as our starting point. The year of Edward Hunter's first book on brainwashing was also the year of publication of Marshall McLuhan's *The Mechanical Bride: Folklore of Industrial Man*. A professor of English literature at the University of Toronto, McLuhan recast himself during that period as an expert on media and mass communication. In *The Mechanical Bride* and later books, he adopted the perspective of an anthropologist studying the "social myths" or "folklore" of tribal society. In the case of contemporary society, he argued, such folklore was largely the creation of the field of public relations. Many of the short entries in his book were accordingly devoted to the car, cigarette, and other ads that formed the visual surround of this new era of material plenty. Ads, he wrote, "are the cave art of the twentieth century."

With its analysis of the efforts of advertising experts to "get inside the collective public mind" in order to "manipulate, exploit, control," McLuhan's book depicted a reality quite as paranoid as that depicted by Edward Hunter, albeit in somewhat more playful terms. "The ad men are constantly," he wrote, "breaking through into the Alice in Wonderland territory behind the looking glass, which is a world of sub-rational impulses and appetites." Like other commentators of the time, McLuhan was fascinated by the way that the modern media sought to "take possession of the nervous patterns of man." He examined closely the methods used by modern advertising and entertainment to address consumers and the way they seemed to plunge them into a trancelike state. "Freedom— American Style" (the title of one of McLuhan's entries) was achieved through a process of conditioning and suggestion that used the methods of Pavlov and Freud to control the unconscious minds of the mass public while preserving the fiction of free will. Both the ad agencies and Hollywood "are always trying to get inside the public mind in order to impose their collective dreams on that inner stage."

McLuhan's writings wove together a wide range of contemporary scientific sources, including Norbert Wiener, J. Z. Young, Margaret Mead, and Hans Selye. Under the influence of Wiener's writings, he began a seminar on culture and communication in 1953. In McLuhan's hands, however, cybernetics took on strange contours. Even as it formed part of the context for McLuhan's analysis of the impact of machines on "human rhythms and social patterns" and their role in imprinting our "spontaneous" impulses, it also seemingly lent itself to forms of magical thinking. Referencing the work of Duke University professor Joseph Banks Rhine on ESP and telepathy, work that had recently received much public notice, McLuhan fantasized about the potential totalitarian implications of such research: "A single mechanical brain, of the sort developed at MIT by Professor Norbert Wiener, when hitched to the telepathic mechanism of Professor Rhine, could tyrannize over the collective consciousness of the race exactly in comic-book and science-fiction style." McLuhan treated pulp genres as worthy of serious attention, according them a value surpassing that of more sober works of prognostication: "The misleading effect of books like George Orwell's *1984* is to project into the future a state of affairs that already exists." In an ad celebrating the toughness of male consumers McLuhan detected deep pathos, for in a world dominated by the abstractions of cybernetics and consumer research, there was increasingly little place for "real men": "The only functions left for the human mind are pure speculation, on one hand, or the manufacture of ever greater mechanical brains, on the other. Those who are not fitted for either of these arduous pursuits—the great majority of men—will inevitably sink into a serfdom for which they have already been very well conditioned."

THOUGH MCLUHAN HIMSELF INCREASINGLY TURNED away from his early gloomy prognostications, his account was echoed later in the decade by the journalist Vance Packard, in whose hands the "paranoid style" that the historian Richard Hofstadter saw entering American public life in the 1950s was wedded to an analysis of the "dark arts" of modern public relations. If the Cold War was often cast as a defense of democratic freedom against Communist unfreedom, Packard's book *The Hidden Persuaders* (1957) suggested that democracy had already been hollowed out by the ever more finely calibrated application of scientific method to the management of mass society. Public life in America, he argued, was increasingly shaped by a scientific caste he called "the depth men," who used a mixture of Pavlov and Freud to work directly on the CNS or unconscious of the consumer-voter. Whereas Hannah Arendt had depicted mass society as a realm in which the ideologies of Nazism and communism contended for "dominion over the minds of modern man"—a process that, she felt, had reached its culmination in the Pavlovian methods used on concentration camp prisoners—Packard offered a no less disturbing vision of modern democracy as a system in which both commercial and political loyalty were captured through techniques similarly modeled on those pioneered by Pavlov to train his dogs. He analyzed in detail the "Orwellian configurations" of the world dreamed up by the persuaders, whose behind-the-scenes operations had resulted in the creation of what one expert cited by Packard called "an unseen dictatorship" that used "the forms of democratic government."

A notable feature of Packard's account was its emphasis on what he called "depth manipulation." There he touched on the larger transformations occurring in the field of behavioral science. The long-standing alliance between commercial advertising and scientific psychology had been cemented early in the century in the relationship between the behaviorist John Watson and the advertising agency J. Walter Thompson. But as we saw in previous chapters, the behaviorist paradigm, which largely ignored the contents of mind and brain, was increasingly superseded in the postwar era by a heightened appreciation of the neural or subconscious factors shaping behavior, motivation, and desire. The depth approach bypassed the rational or conscious parts of the mind, burrowing deeply into its subconscious strata. The new science of persuasion borrowed liberally from both Pavlov and Freud, using techniques of reflex conditioning to induce customers to buy a product by "getting the product story 'etched in his brain'" or the insights of psychoanalysis to perform its work within the "crevices in the human psyche."

Like McLuhan, Packard relied often on the notion of hypnosis in his depiction of the process of training or conditioning Americans to develop loyalty to brands and politicians. He cited scientific studies that pointed to evidence of decreasing rate of eye blinks among supermarket shoppers, seeming to indicate that they had entered into a trance state. He also reported on experiments with subliminal messaging, in which product ads were flashed at split-second intervals during a regular film screening, in such a way that they bypassed the viewer's conscious awareness to penetrate directly into her or his subconscious. Though Packard maintained an attitude of measured skepticism toward such "subthreshold" effects, he nevertheless treated them as indicative of future possibilities.

Toward the end of his book, Packard veered entirely into the realm of speculative science: "Eventually—say by A.D. 2000—perhaps all this depth manipulation of the psychological variety will seem amusingly old-fashioned. By then perhaps the biophysicists will take over with 'biocontrol,' which is depth persuasion carried to its ultimate. Biocontrol is the new science of controlling mental processes, emotional reactions, and sense perceptions by bioelectrical signals." Sketching a future that contained elements of both cybernetics and Huxley's *Brave New World*, he noted that "Planes, missiles, and machine tools already are guided by electronics, and the human brain—being essentially a digital computer—can be, too." Citing the predictions of one contemporary electrical engineer: "The ultimate achievement of biocontrol may be the control of man himself. . . . The controlled subjects would never be permitted to think as individuals. A few months after birth, a surgeon would equip each child with a socket mounted under the scalp and electrodes reaching selected areas of brain tissue. . . . The child's sensory perceptions and muscular activity could be either modified or completely controlled by bio-electric signals radiating from state-controlled transmitters." Like the vatted brain of Siodmak's story, all inputs could in theory be controlled.

ONE OF THE MOST CREATIVE readers—or, as some felt, misreaders—of midcentury brain science was the polymath novelist and philosopher Aldous Huxley. The application of the paranoid style to social commentary reached a kind of apotheosis in Huxley's *Brave New World Revisited*. What had been presented as science fiction in the original version of *Brave New World* was now offered up simply as reportage about a future that had become the present—one that, in Huxley's view, resembled less the nightmare police state of Orwell's *1984* than a more insidious world in which people had been conditioned into embracing their subjugation.

The product of a family of eminent British scientists, Huxley had to abandon his plans of becoming a doctor after an early illness left his vision seriously impaired. Virtually blind by the 1950s, he nevertheless maintained his lifelong interest in science and its social applications while also exploring numerous forms of visionary practice and mysticism. During the early 1950s, having moved to California, he embarked upon a period of sustained self-experimentation with methods of what he called self-transcendence, including meditation, yoga, hypnosis, drugs, and other esoteric practices. He sought out gurus of all kinds, at one point immersing himself in the teachings of Dianetics, though he experienced frustration in his efforts to recapture childhood memories so as to "go clear" (and found Hubbard himself "curiously repellant").* He also famously experimented with hallucinogenic drugs including mescaline and LSD. He wrote about those experiences in two books that became classics of the 1960s counterculture, *The Doors of Perception* (1954) and *Heaven and Hell* (1956). In those and subsequent writings Huxley articulated a theory of the brain as a "reducing valve": a system that filtered the vast streams of stimuli flooding it from all directions—external and internal—in order to maintain optimal efficiency in its functioning. According to Huxley, psychedelics temporarily disabled that filtering system, thus opening the mind to new perceptual experiences, including the cosmic realm of what he called "Mind-at-Large."

Though these books were devoted to the hallucinatory states Huxley experienced while on drugs, they also included mention of other pathways to visionary experience. One of them was Grey Walter's stroboscope, a device whose "rhythmically flashing light seems to act directly, through the optic nerves, on the electrical manifestations of the brain's activity." Huxley noted both the risk of epileptic seizure and the possibilities for potentiating the hallucinatory properties of mescaline and lysergic acid

---

* In a letter to a friend, he reported on his experience with Dianetics: "Up to the present I have proved completely resistant . . . there is a complete shutting off of certain areas of childhood memory, due, no doubt, to what the dianeticians call a 'demon circuit,' an engrammatic command in the nature of 'don't tell,' etc." David King Dunaway, *Huxley in Hollywood* (New York: HarperCollins, 1989), 278.

associated with flicker. In the same passage he noted Wilder Penfield's finding that probing an epileptic patient's exposed brain tissue with an electrode induced the recall of a chain of memories relating to a past experience as well as the emotions aroused by that experience. Huxley mused about the implications of Penfield's research for a deeper understanding of visionary experience: Was it possible, he wondered, for the electrode to elicit William Blake's cherubim? If "visionary experiences enter our consciousness from somewhere 'out there' in the infinity of Mind-at-Large, what sort of an *ad hoc* neurological pattern is created for them by the receiving and transmitting brain?"

But even as Huxley experimented with pathways to personal ("upward") transcendence, he continued to conjure dystopian scenarios associated with the new methods of "downward" transcendence now available to modern despots. Such methods offered, in place of the burdens of modern selfhood, the appeal of immersion in the mass. Among these possibilities was a host of what he called "new and previously undreamed-of devices for exciting mobs." They included radio, movies, and TV. *Brave New World Revisited* included a chapter on brainwashing that discussed those and other means of exercising control over populations. Huxley's original novel had incorporated Pavlov's techniques of conditioning into its account of social engineering, and those techniques were again referenced in his new work, now augmented by the account of Pavlov offered in the British psychiatrist and brainwashing expert William Sargant's *Battle for the Mind: A Physiology of Conversion and Brainwashing* (1957). According to Sargant, the experience of two world wars had confirmed Pavlov's finding that prolonged stress could bring about states of nervous collapse similar to those induced in his laboratory animals. But what had been a by-product of those wars became, in the circumstances of the Cold War, the basis of a methodical, calculated approach to bringing about "cerebral breakdown." Based on his reading of Sargant, Huxley concluded that the interrogation methods used during the Korean War had perfected the use of Pavlovian methods for the process of making prisoners suggestible to the point where they would confess to literally anything. Having been tested in that context, he suggested, they were now ready to be applied on a mass basis.

Huxley predicted that the scientifically trained dictator of the future would equip all schools and hospitals with systems for delivering taped persuasion to students and patients on a continual basis. Indeed, he speculated, in a world facing a dire future of overpopulation and ever-scarcer resources, the practice of brainwashing hitherto associated with police states and prison camps would eventually give way to new forms of government: "It represents the tradition of *1984* on its way to becoming the tradition of *Brave New World*." Such an approach would dispense with brute force and rely instead on the use of drugs and hypnopedia, or sleep training, to influence subjects from infancy onward. In that respect, Huxley's account departed not just from that of Orwell but from the behaviorist model that had provided the underpinnings of *1984*. By dismissing biology, heredity, and brain chemistry, behaviorism, he wrote, fell short of encompassing a full account of human existence. What Huxley called a "tidal wave of biochemical and psychopharmacological research"—he cited the recent discovery of chlorpromazine, Miltown, pentothal, and sodium amytal, as well as LSD—had delivered a veritable cornucopia to modern medicine and by extension to modern rulers. "Pharmacology, biochemistry and neurology are on the march . . . in the course of the next few years, new and better chemical methods for increasing suggestibility and lowering psychological resistance will be discovered." The same held true of new methods of subconscious persuasion such as subliminal messaging, hypnopedia, and so forth. The appearance of such methods would have dire implications for the future: "The ideals of democracy and freedom confront the brute fact of human suggestibility."

That Huxley intended his book as a response to *1984* is made clear in a letter he wrote to Orwell following the publication of the latter's novel. In it he praised Orwell's contribution to what he called a "philosophy of the final revolution," a revolution that aimed at the "total subversion of the individual's psychology and physiology." His letter went on to express doubts, however, about Orwell's emphasis on the brute force of totalitarian methods and suggested that future systems would employ techniques of rule more closely resembling those of *Brave New World*. Placing special emphasis on the political significance of hypnotism in the modern world, he wrote that in combination with barbiturates and methods of psychoanalysis, states of heightened suggestibility could be manufactured in even the most recalcitrant subjects: "Within the next generation I believe that the world's rulers will discover that infant conditioning and narco-hypnosis are more efficient, as instruments of government, than clubs and prisons, and that the lust for power can be just as completely satisfied by suggesting people into loving their servitude as by flogging and kicking them into obedience."* In that respect Huxley's speculations about a "Manhattan Project" of human happiness were considerably darker than Orwell's dystopia. In his view, the heightened states of hallucinatory suggestibility common to POWs and other individuals subject to total "milieu control" would become a feature not just of totalitarian societies but of purportedly democratic societies as well.

Huxley's dramatic claims did not go unchallenged. The British psychiatrist J.A.C. Brown, for one, described them as wildly overblown. He traced those claims to a basic fallacy. Huxley, like Sargant, he said, supposed "that mind and brain are synonymous," whereas, he argued, although the mind might be located in the brain, it was not reducible to the brain: "Mind is a social product." He cited a passage in Sargant's book that touched on Grey Walter's discovery that strobe could produce abnormalities of brain function, including the symptoms of epilepsy—hallucinations, intense emotions, disturbances of the sense of time—even in healthy subjects. He went on, however, to add that such responses seemed to die away with continuous exposure, suggesting that the brain "'learned' not to respond in an abnormal way." Indeed, he noted, such experiences could apparently be controlled by the subject, once again citing Walter: "The will of the subject can also be brought into play: he can, for instance, consciously and with effect resist or give way to the hallucinations engendered by the flicker, a matter of no small social interest as well as enlightenment on the question of self-discipline." Brown likewise

---

* Ewen Cameron's experiments in "depatterning," discussed in previous chapters, were apparently inspired by Huxley's novel and seemed to give brute form to Huxley's predictions concerning the "final revolution" of physiology and psychology.

considered the evidence for the assertion that people could be placed in deep trance states against their will flimsy. As for Huxley's warnings concerning the role of the mass media in exciting mobs, Brown pointed out that that was hardly something new in history but merely reproduced the states of religious excitement common to crowds in the Middle Ages. Such views, however, did little to stem the tide of paranoid commentary on the mass media. Whether it was the dangers of subliminal messaging (which studies showed to be largely ineffective) or flicker (such as the anecdote of the moviegoer who tried to strangle his seatmate under the influence of the flickering image on-screen), such dangers became part of what McLuhan called the "folklore" of modern society: a vernacular understanding of media effects whose emergence accompanied the ever more pervasive spread of those effects.

Dawn of the Televisual Era

A good illustration of how the midcentury paranoid style found expression in popular culture is the 1962 remake of *Donovan's Brain* (1953). That version was simply titled *The Brain*, and the story it told is as follows: After the tycoon Holt's death in a suspicious plane crash, the doctor Patrick Cory steals his brain, keeping it alive in a vat and studying its signals with the help of the ever-present EEG. Soon the envatted brain acquires power over Cory. But whereas in the original Cory becomes an instrument of a criminal conspiracy orchestrated by the brain, in the remake he is enlisted by the brain to uncover the plot to murder Holt. In the final scene it is revealed that the plot originated with Holt's daughter, who wants to stop her father's efforts to impose monopolistic control over a new miracle drug that has the potential to cure millions.*

The remake adds fresh elements to the original's portrayal of a world permeated by technological devices. Numerous transcribing, recording, and lighting systems figure in the story. First and foremost, there is the

---

* This drama may represent the original brain-in-a-vat scenario later referenced by the philosopher Hilary Putnam.

EEG machine itself, endlessly churning out its "slow, trembling curves."
Another technological device that features prominently in the remake
is the stroboscopic lamp. At key moments in the story, strobe seems to
take on a life of its own. Its mysterious triggering in one scene induces a
fatal seizure in a character who tries to steal the brain. A similar lamp is
used in a later interrogation scene, while a scene in a nightclub features a
quasi-stroboscopic disco ball. And it is the stroboscope in Cory's laboratory that eventually kills off the brain by triggering a seizure—this time
in the brain itself—and thus releasing the doctor from its spell. The role
played by the device evokes strobe's increasingly ubiquitous presence in
midcentury society. Pediatricians of that era noted with concern the per-
vasiveness of flicker in the modern world (in fluorescent lighting, movies,
and TV) and warned those prone to seizures to exercise due caution.

The film's vision of a technologized world includes many uncanny
and magical elements, among them the telepathic form of communication that links the brain to Cory. Brain waves, it seems, correlate with
other kinds of invisible signal patterns that permeate the atmosphere,
exercising a mysterious influence over all who enter their range.* Unlike
the original film adaptation of Siodmak's novel in 1953, the 1962 remake is marked by its incorporation of TV as a thematic element. As in
the case of the stroboscope, television is endowed with mysterious agency.
TV sets appear on-screen several times, most notably in a scene in which
Cory tracks down a man he wants to question in relation to Holt's death.
Upon arrival he finds the man watching a western on TV. At a key moment the viewer sees, in a close-up, a pistol pointing out of the TV set,
whose screen briefly fills the entire movie screen. Cory, who has been
momentarily called out of the room to answer the telephone, returns to
find that the owner of the TV has been shot dead. The scene comments
directly on the midcentury cultural panic surrounding the mass media
and its perceived hold over audiences.

---

* In the adaptation released in 1953, Cory cited the experiments of Duke University's
Joseph Rhine that McLuhan had mentioned in connection with the "telepathic" possibilities of Wiener's mechanical brain.

As the film hints, there is a curious parallel history uniting the TV and EEG. Both were invented in the late 1920s and came into their own in the post–World War II era, and these developments were linked in a number of ways. The cathode ray tube, which translates electronic signals into visual images, became central to early television design and was likewise adapted by Grey Walter in the design of his toposcope, a multichannel recording device for reading brain waves and projecting them onto a screen, rather than, as had hitherto been the custom, transcribing them via ink onto paper (the new system was housed in a laboratory whose appearance one visitor to Walter's clinic likened to a TV studio). Walter himself underscored the parallels between the two in suggesting that the televisual system functioned like the neurovisual one, in other words that visual perception functioned through a scanning mechanism similar to that of television. *The Brain* alluded to that by portraying the brain in its tank, endlessly scanning the surrounding environment for signals. At the same time, however, Walter shared the concern of many that excessive TV viewing might adversely affect the developing brain (and tried to measure that with EEG). As the fifties transitioned into the sixties, it was the dangers of the new medium for viewers, especially but not only those who were underage or seizure prone, that aroused concern.

Television became a massive presence in American society in the 1950s. From 1950 to 1954, the number of TV sets owned by Americans shot up from 4.4 million to 44 million; as a percentage of American homes, TV ownership rose from 11 percent to 88 percent. Hundreds of millions of individuals began spending many hours sitting "in close proximity to flickering, light-emitting objects." By 1960, the year of the Richard Nixon–John F. Kennedy presidential debates, Americans were averaging four to five hours of TV watching per day.

The staggering dimensions of this development generated widespread commentary. The journalist Theodore H. White wrote of the "blast effect of this explosion on American culture" and expressed alarm about TV's power to "shape the American mind." The medium's penetration of the private space of the home generated particular unease, especially among those who shared the view of the philosopher Hannah Arendt concerning the vital nature of the private realm for the qualities associated with citizenship. The advent of TV seemed to some to point to a future in which "citizenship was supplanted by viewership." Exposed for hours at a time to the saturation messaging of commercial advertising, Packard wrote, audiences underwent a process of deep conditioning. The process was carried to still greater extremes in the watershed political campaigns of 1956 and 1960, when the voter, according to Packard, was treated like Pavlov's dog and many observers, including Kennedy, credited TV with playing a decisive role in the election outcomes. The captive audiences sought by America's TV industry seemed to replicate the condition that, according to Eastern European dissidents such as Czesław Miłosz, had molded the "captive minds" of the Soviet bloc. Struck by such parallels, some Western social critics began to identify captivity as a condition increasingly common to both Communist and capitalist societies. The widespread belief in the West that the public was free from the relentless messaging campaigns to which citizens behind the Iron Curtain were subjected, which Lajos Ruff had described as part of an extensive "treatment" in a "magic room of Communist propaganda and ideology," was in that sense simply another of the self-flattering delusions of democratic society.

Messaging aside, did the televisual medium possess other properties that extended its hold over audiences? It was McLuhan who went farthest in proposing answers to that question. According to the Canadian sage, the singular feature of the TV image was its lower definition vis-à-vis the film image, a quality that made TV viewing a more intensely immersive experience and one that demanded the audience's correspondingly greater involvement in the process of "filling in" missing information. Mused McLuhan, "What possible immunity can there be from the subliminal operation of a new medium like TV?" Here he was alluding to the specter of subthreshold effects that advertisers had experimented with in the late 1950s. The furor that surrounded subliminal messaging crystallized the anxieties about uncanny agency that observers at the beginning of the decade had expressed in relation to the mass media more generally. In that regard, McLuhan argued, TV went far beyond movies. As competition between the two media intensified, the place of TV in American life accordingly became a subject of deep concern. Anxieties about the film industry's role in implanting subversive messaging in American citizens, which came to a head in the House Un-American Activities Committee (HUAC) investigations into Hollywood at the beginning of the 1950s, were eclipsed by the paranoia that came to surround TV.

The Manchurian Candidate

That paranoia is central to the plot of John Frankenheimer's *The Manchurian Candidate* (1962), adapted from Richard Condon's 1959 novel. In this dark satire, the variations woven by Hollywood on the theme of mind control reached a delirious climax. Though the film tapped deeply into Cold War paranoia, it also turned it inside out, collapsing mind control Communist and capitalist style into each other and ultimately rendering them indistinguishable. The mass media, and in particular TV, became crucial to that reversal. Yet though the film went to great lengths to implicate TV in its depiction of a homegrown version of mind control and to align film with a more critical, self-aware representation of reality, Frankenheimer did not entirely succeed in his efforts to distance cinema from the anxieties that pervaded midcentury views of the mass media. In the immediate aftermath of JFK's assassination in 1963, Richard Condon wrote in the *Nation* that Americans had been relentlessly "conditioned to violence" by what he called "the over-communications industry" and suggested that there were many potential Manchurian Candidates at loose in American society. The film was withdrawn from circulation soon thereafter, leading to speculation that this had occurred in direct response to a perceived connection between the film and the president's murder.

The figure at the center of the story, Major Raymond Shaw, is a decorated war hero who returns from Korea to receive a Medal of Honor. He becomes the object of a fawning press, whipped on by his scheming mother, Eleanor, and his father-in-law, Johnny Iselin, a right-wing blowhard senator. Following a script prepared by Eleanor, the Joseph McCarthy–like Iselin begins hurling allegations at the army that it is harboring Communists, while seizing every opportunity to be photographed with his war hero son-in-law. Yet Shaw's hero narrative is a

fiction masking a sinister reality: the firefight in which he performed
the actions for which he received his medal had been a charade, part of
a carefully orchestrated scheme in which he and the men of his unit had
been taken prisoner, flown to a location in Manchuria, and then worked
over by scientists at the Pavlov Institute under the supervision of its director, Yen Lo. Their testimony about the imaginary firefight is based on
a carefully fabricated and then implanted memory.

Condon's novel treated the brainwashing process in overtly cinematic
terms. Now back in the United States and working for army intelligence,
the captain of Shaw's unit, Major Bennett Marco, begins to suffer nightmares relating to the incident in Korea. The nightmares call into question
significant elements of the official story about the event. Marco's dreams
include increasingly disturbing details, among them a projection room in
which the members of his unit were subjected to memory drills following
their capture: "He thought about the movement of the many red dots on
the screen, then of Raymond, symbolized by the blue dot, and the canned
voice telling them they were seeing the battle action in which Raymond
had been willing to sacrifice his life, again and again, to save them all."
He questions Raymond about the incident, but Shaw remembers nothing
about memory drills or a movie projection room—a lack of recall that
Marco eventually interprets as evidence of the fact that he underwent a
more intensive process of brainwashing.

In fact, as the reader of the novel has by this point already learned,
Yen Lo's staff has trained the members of the unit in four separate versions of the imaginary engagement, each representing a different vantage
point: "Each patrol member had been drilled in individual small details
of what Raymond had done to save their lives." In the case of Raymond,
Yen Lo had personally taken charge of the process, "utilizing a pioneer development of induced autoscopic hallucination which allowed Raymond
to believe he had seen his own body image projected in visual space." The
passage suggests that the process of implanting a false memory that occurs in brainwashing is akin to that of crafting a film scenario—that the
brainwasher is, among other things, a master of such scenarios. As the
story unfolds, the line between memory, hallucination, and film becomes
increasingly blurred.

The end result is a lesson in the fact that, as Yen Lo states in the novel's most telling line, "The first thing a human being is loyal to . . . is his own conditioned nervous system." As the head of the so-called Pavlov Institute, Yen Lo proceeds to establish the reality of brainwashing for the benefit of an audience of Chinese and Soviet officials gathered in the institute's Research Pavilion. In doing so he refers to several of the authorities cited in *KUBARK*, the CIA's in-house interrogation manual that was finalized the same year that the film adaptation of Condon's novel was released. Yen Lo begins by dismissing "the old wives' tale" that hypnotized subjects cannot be forced to do something against their will by citing from an extensive Western scientific literature, including works on hypnotic crime, as well as on conditioning through comic books (here he cites the work of Fredric Wertham—himself, as we saw, a student of Pavlov's work). Throughout the story he repeatedly and slyly also invokes the power of public relations to influence and mold individual behavior, ironically paraphrasing cigarette ads to his less-clued-in Soviet assistants. In winking at the possibility that the cigarettes smoked by his captives may have been "doctored" (with yak dung), he seems to allude to the "magic room" cited in *KUBARK* in which cigarettes whose flavor has been modified are used to convince the subject that he has been placed in a trance (see chapter 4).

But in Frankenheimer's film, visual proof is required for Yen Lo's audience of skeptical Soviet apparatchiks. This is supplied in the famous garden club scene. The scene is a cinematic tour de force that employs an extended pan to stitch together in the same visual frame two quite incompatible realities: in one the audience sees things from the perspective of Shaw and his men, who believe themselves to be in a hotel in New Jersey attending a meeting of the local ladies' garden club; in the other, which is revealed to us as the film camera continues its 360-degree pan, Shaw and his men are shown to be seated in a trance state on the stage of an amphitheater, whose tiered seats are filled with Chinese and Russian officials. Confirmation that hypnotic crime is no mere old wives' tale is supplied when Shaw, acting on Yen Lo's orders, murders one of his own men by strangling him. In a subsequent version of the scene, presented to us as a nightmare dreamt by one of the surviving men of the unit, Shaw is shown shooting another of his men. As in the aforementioned film *The Brain*, the scene is shot in such a way as to comment on the perceived influence of the media over human behavior: Shaw's gun points directly out of the screen at us. In the haunted dreams of the soldier, the real scene has begun to bleed through the false memory of the implanted scene.

The most purely cinematic moment in *The Manchurian Candidate*, the garden club scene also represents the purest mass cultural representation of the syndrome known as the "prisoner's cinema": the hallucinatory state that is elicited by specialized scientific techniques. Yet although the scene implicates film in a generalized panic about the malign agency of the mass media, in subsequent scenes such anxieties are displaced onto its rival television, which during the 1950s had taken the place of movies as the primary form of mass entertainment in American society. American-style mind control via television emerges as a central plot theme in Frankenheimer's film, even as the hallucinatory state experienced by Shaw and his men is projected onto the national stage. *The Manchurian Candidate* thus masked the uneasy history of HUAC's investigation into Communist subversion in the film industry by aligning television, film's rival, with a far greater homegrown threat operating on the opposite end of the political spectrum: that of the histrionically anti-Communist Right.

This is dramatized in a scene in which Johnny Iselin announces his bombshell charges to a stunned Senate chamber. In the foreground of the scene, the viewer sees a television monitor transmitting a live broadcast of the proceedings unfolding in the background. The close affinity between the political pathology associated with anticommunism and the televisual medium is here made manifest. Raymond's mother, Eleanor—the real power behind the throne—is shown hovering over the monitor in a way that serves to drive home the point that it is she who has scripted her husband's lines and scene as well as the larger strategy they set in motion. The strategy is later spelled out by Mrs. Iselin when, in the film's second major plot twist, she reveals herself to be Raymond's real programmer by speaking the trigger phrase that places him in a trance: "Raymond, why don't you pass the time by playing a little solitaire?" Obeying his mother's suggestion, Raymond quickly turns up the queen of diamonds, thus unlocking the mechanism that allows her, now revealed as the *Alice in Wonderland*–like Red Queen, to give him his instructions for the assassination that he is meant to carry out at the Republican National Convention—an assassination intended to "rally a nation of TV viewers into hysteria" and clear the way for her buffoonish husband to assume the presidency.

John Frankenheimer, The Manchurian Candidate *(1962).*

Though it is the ideological battleground of the Cold War that provides the film with its brainwashing scenario, the film also increasingly implicates the mass media in the wider phenomenon of mind control. The medium itself, as McLuhan would have it, is the message. The delivery of that message occurs in stages: in the movie's first half it is film, but in the second half it is the "blast effect" of TV that serves as the thematic focus of its treatment of the conditioning of individuals, audiences, and, ultimately, the entire nation. In the end TV's power to capture mass audiences is depicted as virtually limitless.* The brainwasher is, again, a maker of potent imagery, only now televisual. Whereas the garden club

---

* According to Theodore H. White, writing about the audience for the televised 1960 debates, "No larger assembly of human beings, their minds focused on one problem, has ever happened in human history." White, *The Making of the President 1960* (New York: Atheneum, 1961), 222.

scene focuses on a handful of men, the convention hall scene—another of the KUBARKian "magic rooms" that proliferate throughout the film—addresses a mass public. In the former scene, Shaw's gun points directly at the audience, while in the latter he is positioned in a spotlight booth with his rifle pointing at the stage in preparation for the planned assassination. We are, as it were, on the other side of the weapon: we adopt the shooter's perspective as he sights the target with his crosshairs.

The distinction to be drawn here lies in the fact that the garden club scene, the most cinematic in the movie, aligns the hallucination with the prisoners. The scene positions us in such a way that we can scan it; within the framework of the film, movies still preserve a relation with truth, albeit a complicated truth that includes recognition of the medium's own deceptive powers. In the convention hall scene, however, such a relationship is impossible. We have been subsumed within the nation of hysterical TV viewers, and it is hardly possible for us to scan "reality." The pleasure we take in the fact that Raymond ultimately thwarts the conspiracy by assassinating his mother and father-in-law can't be separated from the realization that the film has taught us to recognize the degree to which this pleasurable reaction is simply one of the conditioned responses continually elicited by the mass media. Every member of the audience is in this sense a potential Manchurian Candidate, as Richard Condon suggested. The transformation of the syndrome known as the "prisoner's cinema," which emerged as part of a tradition of mind control experimentation that culminated in the contemporaneous CIA document KUBARK, into an element of the critique of Western modernity was accompanied by the realization that that syndrome, manifest in the amnesiac, hallucinating condition of Raymond Shaw, was the normal condition of the modern self.

Coda

The phantasmal danger projected onto the mass media at midcentury seemed to find real-world expression when The Manchurian Candidate was pulled from circulation following the assassination of JFK (though the reasons for this remain murky). Among the speculations concerning the actions of Lee Harvey Oswald was the possibility that he had seen and been inspired by Frankenheimer's film to carry out his murderous deed. In the months before the assassination, it was subsequently shown, he had regularly passed a theater exhibiting the film on his way to work. Further investigation turned up the fact that it may simply have been one of Oswald's several "trigger films."

Brainwashing narratives, Frankenheimer's film shows, were not simply a product of Cold War tensions; they also functioned as part of a vernacular understanding of the media effects that became pervasive in the postwar era. Those effects, as we shall see in the concluding chapters, were conceived of as operating in one of two ways. Oscillating between what McLuhan called hot and cool media, midcentury audiences veered between over- and understimulation. As the historian Jimena Canales wrote in relation to McLuhan, a hot medium such as film creates effects similar to those produced by flicker: the overactivation (or, in McLuhan's terms, the "hotting-up") of one sense. A cool medium such as television, by contrast, generates effects similar to those produced by sensory deprivation: in experiments in which all outer sensation is withdrawn, wrote McLuhan, "the subject begins a furious fill-in or completion of senses that is sheer hallucination."

CHAPTER 6

Brain Science in the Visionary Mode

Brion Gysin, William S. Burroughs and the Dream Machine.

Recent medical incidents in Britain have shown that
flickering light can produce epileptic fits, even in people who
are not prone to epilepsy. "Photic epilepsy," as it is known, is
caused by a source of light pulsing at exactly the same rate as
human brain impulses—even watching television has been
known to bring it on. William Burroughs describes here what
the incredible results of this phenomenon could be if taken
to its logical conclusion: a method by which completely new
characteristics can be projected on to a man's body and mind.

—"SWITCH ON AND BE YOUR OWN HERO,"

THE BURROUGHS ACADEMY, BULLETIN 9, IN MAYFAIR, JUNE 1968

It was man's "tardy recognition of the organ that alone sets him apart from other living beings" that, according to the neurophysiologist Grey Walter, marked the great accomplishment of mid-twentieth-century brain science. The moderns had now amended the ancients' injunction "know thyself" with their new exhortation to "know your brain." But what exactly did it mean to adopt a view of man that placed the brain, the most refractory of the body's organs, at its center?

Earlier chapters of this book traced the relation of midcentury brain science to many of the deepest anxieties of the 1950s: to the ethical and philosophical qualms surrounding research on the brain; to the Cold War specter of mind control; and to mounting concern about the impact of the mass media on human behavior, memory, and identity. As the 1950s transitioned into the 1960s, those anxieties grew ever sharper, shaping central elements in the worldview of the emerging counterculture. This term is a notoriously slippery one comprising many different groupings and tendencies. Here it is used primarily in relation to a form of literary and cultural experimentation associated with the Beat movement and especially with William S. Burroughs, one of the essential figures of the counterculture, yet one who remained in many ways an outlier to it. Burroughs's relationship to midcentury brain science, particularly to the strand of it associated with Grey Walter, is the focus of this chapter. Though it later became common to claim, as did the historian Theodore Roszak in *The Making of a Counter Culture: Reflections on the Technocratic Society and Its Youthful Opposition* (1969), that the counterculture was largely hostile to science, this is wrong on many counts.* Not least among the reasons is the way it ignores the many connections between midcentury brain science and the explorations undertaken by that culture. In certain respects the counterculture may be seen as another of the mutant offspring of the scientific inquiries traced in previous chapters. This is especially true in relation to psychedelic drugs—about which much has been written concerning the connections between the 1950s and 1960s—but also of flicker and sensory deprivation, about which much less has been written. The final sections of this book are devoted to these latter two phenomena.

THE NEW PICTURE OF THE brain that emerged at midcentury opened many unfamiliar vistas. They included treatments for maladies such as epilepsy, an illness that since ancient times had been seen as a form of "sacred possession" but that now began to yield its secrets to modern medicine. Answers to old questions about the nature of consciousness, perception, and memory seemed within reach. Life itself became redefined in relation to the brain: previously, the heart had been seen as the vital organ, whose loss of functioning signaled the end of life, but in the 1960s, brain death, as measured by the cessation of EEG activity, was established as the event marking the end of life. The EEG's rhythmic oscillations became recognized as "a stand-in for life itself."

But if midcentury brain science awakened new hopes, it also inspired

---

*   "If the resistance of the counter culture fails, I think there will be nothing in store for us but what anti-utopians like Huxley and Orwell have forecast." Theodore Roszak, *The Making of a Counter Culture: Reflections on the Technocratic Society and Its Youthful Opposition* (New York: Doubleday, 1969), xiii.

anxiety, even terror. The prospect of brain studying brain—of making the organ of knowledge its own object of knowledge—produced a kind of vertigo. One place to look for the fascination and unease evoked by those developments is in horror and science fiction. Norbert Wiener's science fiction story "The Brain" (1952) captured this sense of unease well. The story concerns a gathering of scientists who fall into a discussion about lobotomy. One member of the group, the director of a local mental asylum, has brought along a patient of his who suffers from amnesia. The morbid discussion triggers a crisis in the patient, and the doctor is forced to give him a barbiturate. Under the influence of the sedative, he experiences a sudden recovery of memory. We learn that he was formerly a brain surgeon who lost his young son following a head injury caused by a speeding car, which was driven by men who work for a local crime boss known as "The Brain." The crime boss then suffered a head injury of his own in a later car accident. The surgeon was brought in to save him, but he exacted his revenge by performing a lobotomy on The Brain. He then suffered a breakdown, descending into the amnesiac state that landed him in the asylum.

Wiener's little tale serves as a kind of parable of the new era. Within the space of a few pages, brains, and the injuries they may suffer, multiply vertiginously. Grey Walter argued that with brain studying brain, man learned to see himself in a mirror—but a better analogy, as his friend Wiener seemed to suggest, might be a wilderness of mirrors. How to respond to the resulting confusion was a question addressed by many contemporaries. Consider, for instance, the way that Wiener's "electronic brain" loomed over Marshall McLuhan's analyses of the ad industry's endeavors to get inside and manipulate the public mind. In both method and content, the field of public relations aimed at an increasing standardization and mechanization of thought and behavior, a process reflected in the electronic brain's emergence. Could that process, wondered McLuhan, be reversed? Would it be possible to assist the public to "observe consciously the drama which is intended to operate upon it unconsciously"? That was the question animating his analysis of modern advertising. For guidance he turned to Edgar Allan Poe's story

"A Descent into the Maelstrom," in which a shipwrecked sailor saves himself by studying and cooperating with the action of the whirlpool that destroyed his ship. McLuhan suggested that, once arrested for contemplation, that "whirling phantasmagoria" might be understood and counteracted.

Grey Walter suggested something similar in stressing the constant flux characterizing brain activity that his research had revealed. Fascinated by the "whirling spiral" that many people saw under flicker, he identified it as a basic principle of selfhood, echoing his friend Wiener: "We are but whirlpools in a river of ever-flowing water . . . not stuff that abides, but patterns that perpetuate themselves." But Walter also suggested that study of those patterns could yield new insight, even the possibility of undoing negative patterns. Like the reversal in Poe's whirlpool scene, Walter's suggestion inspired many of the experiments of the 1960s. As we shall see, the specter of brainwashing, though it inspired terror, also pointed to alternate possibilities of counterbrainwashing, of deconditioning, of what the counterculture called "expanded consciousness." Emphasizing those conjoined possibilities brings us back once again to the problem of treating brainwashing as myth. To do so is to miss the extent to which that notion became part of both a vernacular understanding of the operations of Western society and the countercultural project of imaging alternatives.*

---

* There are few better illustrations of this than a scene that is part of the folklore of the counterculture. In 1959, the poet Allen Ginsberg volunteered for one of Stanford's CIA-sponsored LSD trials held as part of an investigation into the drug's use as a "truth serum." At Burroughs's urging, Ginsberg asked the researchers to use a stroboscope when administering his EEG tests. The combination of psychedelic drugs and rhythmically flashing lights produced remarkable effects: "It was like watching my own inner organism," reported Ginsberg. "Suddenly I got this uncanny sense that I was really no different than all of this technical machinery around me." The experience soon turned nightmarish: "I thought I was trapped in a giant web or network of forces beyond my control that were perhaps experimenting with me." Martin Lee and Bruce Shlain, *Acid Dreams: The CIA, LSD, and the Sixties Rebellion* (New York: Grove Press, 1985), 59. From the crisis precipitated by that experience Ginsberg emerged, paradoxically, as a prophet of a heightened consciousness completely liberated from external controls.

Encephalic Civilization and Its Discontents

Grey Walter was one of those who showed the way. Though parts of *The Living Brain* struck the technocratic note that Roszak identified as central to countercultural critique, Walter was no mere engineer of souls, lacking one himself. In his personal life and in his experiments and writings on topics ranging from artificial brains to epilepsy, madness, hallucination, and yogic practices, he charted much of the terrain later explored during the 1960s. Walter also dabbled in science fiction. His novel *The Curve of the Snowflake* (1955) featured time travel, a cybernetic society of the future, and an EEG specialist named Wing. When one character encounters a strange spacecraft, Wing initially suggests that it may be a hallucination and relates such visionary experiences to epilepsy and migraine. He explains it as a disturbance of the field of vision that produces darkness but that also possesses a "light of its own" capable of producing visions such as those experienced by William Blake or the kind Alice had when she went through the looking glass. Such visions, he continues, are not indicative of brain damage or an abnormal EEG reading: "You have quite rich ones with flicker." The vision of the spacecraft ultimately turns out to hold the key to the society of the future envisioned in the novel.

In the preface to the first edition of *The Living Brain* (1953), Walter strove to reassure readers disturbed by the "risky" nature of his subject matter. After he gave a series of talks on the BBC, he wrote, some listeners reported that they had felt a sense of indecency in the idea of brain studying brain, "as if suddenly coming upon themselves for the first time in a looking-glass." Such feelings were misplaced, for, as he put it, "Our peeping here is as innocent as Alice and kinder than Analysis." Though keen to introduce his readers to recent findings in his field, he took pains to profess that he maintained a "reverent attitude to Man." If the suspicion of "peeping" clung to his book, it was that of Lewis Carroll rather than of Sigmund Freud. Walter was not alone in his reverence for Carroll, who makes repeated appearances in the pages of the book. Indeed, Carroll was adopted as a kind of patron saint of midcentury brain science; references to Alice abound in the scientific writings of the era, repeatedly invoked by researchers attempting to identify the strange, even fantastical nature of their findings with a literary tradition seemingly untainted by the scandalous notions of the Freudians.

Yet it was not just the air of impropriety from which Walter sought to distance his book. In its materialistic account of the brain, the picture it drew was a disturbing one, as Wiener seemed to recognize more clearly than he. Midcentury brain science deeply unsettled traditional beliefs about selfhood. During the decade that elapsed between the two editions of Walter's book (1953 and 1963), medical advances occurring on many fronts led to what has been called a vast "enlargement of the field of the therapeutic." New treatments for diseases such as epilepsy, schizophrenia, depression, and migraine and new paradigms of care and well-being were established. The implications of those developments were especially consequential in the field of psychiatry, which was transformed by the discovery of chlorpromazine and an ever-expanding host of other drugs in the early 1950s. Yet although this new psychopharmacological cornucopia stimulated psychiatrists to dream of a future unburdened by the errors of the profession's prescientific past, the all-too-recent history of the age of "heroic treatments" continued to cast a long shadow over that future. The 1950s' revolution in psychiatric treatment was haunted by the shock therapies and aggressive somatic treatments first developed in the 1930s and '40s. That decade of dramatic progress was also, as noted in earlier chapters, the decade of the birth of antipsychiatry, a movement prone to challenging all claims of progress in modern medicine.

In the preface to the second edition of his book (1963), Walter paid tribute to the many advances that had occurred over the previous decade, noting that his method of studying epilepsy by means of flicker was now practiced around the world. Yet he again felt obliged to offer reassurances to his readers. Now, however, the context had changed significantly. It was less religious views and more the geopolitical tensions of the Cold War that shaped the public reception of brain science. With Huxley's *Brave New World Revisited* probably in mind, he reflected, "With increasing understanding comes growing power, and there are some who already see in our tentative essays the means for effective thought-control. . . . To all such we say, as to politicians, preachers and psychiatrists: let only those whose hands are clean busy themselves with brain-washing."

Yet that disavowal was not a little disingenuous, given Walter's own endorsement of the methods of what he called the "heroic physical treatment of mental disorder." Those methods were central to the attacks of the antipsychiatry movement. To antipsychiatrists such as R. D. Laing, the new brain paradigm was doubly suspect: its perceived reductivism and its aggressive somatic methods were entirely of a piece with each other. By the midsixties Walter was himself striking a new note that in some respects echoed the stance of Laing. He now viewed the brain as less a biological than a social organ, and he stressed that the new forms of self-knowledge opened up by study of the brain must include awareness of the myriad threats facing that organ in the modern world, among them all forms of interference with brain processes, including propaganda, drugs, and "mental health" campaigns. By that time, liberation from such forms of "interference" had become a shibboleth of the counterculture in its many forms. For the members of the emerging cultural vanguard and the antipsychiatry movement, brainwashing was simultaneously a nightmare of modernity and a kind of portal offering escape from that nightmare, a pathway to the altered states that so fascinated Walter, such as the looking glass that his beloved Alice went through.

Visions in the Alpha Band

An irony of the Walter story is that although his book *The Living Brain* warned readers about gurus such as L. Ron Hubbard—whose popularity, Walter suggested, was based on his naive valorization of the power of mind over brain—by the end of the 1950s, Walter himself had become a near-gurulike figure among parts of the emerging counterculture. Traces of his influence can be found in many places: as one scholar puts it, behind writers such as Aldous Huxley and William S. Burroughs, you find Grey Walter; behind antipsychiatrists such as R. D. Laing and Gregory Bateson, you again find Walter. Aldous Huxley was only the first of many to make Walter's discoveries about flicker part of his exploration of visionary experience. By the 1960s, the method's uses in opening pathways to altered states of consciousness had come to be appreciated by many. Of none was that more true than William Burroughs and the circle surrounding him, especially Brion Gysin and Ian Sommerville. Walter became Burroughs's guide to the glad new tidings of contemporary brain science, whose advances Burroughs followed closely even as he immersed himself in the doctrines of Scientology. In his correspondence from that period, Burroughs underscored the irony noted above by referring on more than one occasion to Walter and Hubbard in the same breath, while also citing both as being superior to Freud. Burroughs's aversion to psychoanalysis was, it has been suggested, rooted in what he saw as its failure to take more directly into account the material basis of consciousness.*

Burroughs originally came across a copy of *The Living Brain* at a secondhand book stall near his Paris hotel sometime toward the end of the 1950s. The following year, during a stay in London, he attended a lecture by Walter, and in October 1960, he informed his friend Brion Gysin of his plans to visit Walter in Bristol for a "flicker date." Writing to Walter to request an appointment, Burroughs explained that he was interested in flicker for two reasons. The first was "its possible therapeutic application in drug addiction" (he referred to an article he was writing on the subject together with a London physician); the second was "the effect of flicker on the creative process." It is not clear what Walter's response was, if any, or whether the "date" ever occurred, yet that marked the beginning of a period of intense interest on Burroughs's part in the possibilities of Walter's method. He regarded Walter's discovery that flicker appeared to break down normal barriers between different brain regions as one of immense significance.

In a letter to Allen Ginsberg later that same year, written while staying with Ian Sommerville in England, Burroughs mentioned flicker again, this time in connection with psychoanalysis, describing the methods of the latter as obsolete: "When I think of the time and money wasted on the preposterous fraud of analysis and what they are paid to do nothing and determined to go on doing nothing." Then he asked, "Have they

---

* Western psychiatry, felt Burroughs, should have developed along "the lines of Pavlov." David Seed, *Brainwashing: The Fictions of Mind Control* (Kent, OH: Kent State University Press, 2004), 137.

picked up on the encephalographic research of Grey Walter? He can remove so called hallucinations by direct brain intervention. Have they picked up on scientology? On the new hallucinogens? All they want is to sit on their fifty-dollar ass." In subsequent letters he again connected Walter and Hubbard, asking whether the "needle action on the E-Meter [the device used in Scientology to "audit" subjects] corresponds to encephalographic action. . . . I wonder if what [the Scientologists] call a floating needle corresponds to alpha waves?"

By the mid-1960s, however, his tone was changing. Though he continued to find the E-meter useful for "pinpointing psychic areas and eliminating interference," he had become deeply suspicious of Hubbard's authoritarian tendencies. He now referred to him as a fascist and connected his teachings to brainwashing: "I think that Scientology is an officially sponsored experiment to test and develop techniques of thought control." In one letter to a filmmaker acquaintance, he proposed a scenario for a science fiction film that would tell "the story of scientology and their [sic] attempt to take over the planet": "Your planet has been invaded and the landing field for this invasion was precisely the human body. The rightness center [in Burroughs's cosmos, the brain region associated with squareness, straightness, having the "right opinion"] in the human nervous system was invaded and taken over by a parasite." The film would be accompanied by a warning that it would be dangerous to watch, hence "only suitable for those with strong nerves and self-reliant character."

In addition to his proposal to conduct experiments combining the E-meter with EEG, Burroughs discussed his plans to try it in combination with another device that had captured his interest. That was the so-called Dream Machine, an invention of Burroughs and Gysin's mutual friend Ian Sommerville, who was studying mathematics at Cambridge and who became Burroughs's lover. Modeled on Walter's experiments with strobe, the device (originally called a flicker machine) consisted of a turntable, light source, and perforated cylinder. When activated, the rotating light source created flicker effects that reproduced alpha rhythms in the eight-to-thirteen-cycles-per-second range. In February 1960, Sommerville described his invention in a letter to Gysin: "I have made a simple flicker machine; a slotted cardboard cylinder which turns on a gramophone at 78 rpm with a lightbulb inside. You look at it with your eyes shut and the flicker plays over your eyelids. Visions start with a kaleidoscope of colors on a plane in front of the eyes and gradually become more complex and beautiful, breaking like surf on a shore until whole patterns of color are pounding to get in. After a while the visions were permanently behind my eyes and I was in the middle of the whole scene with limitless patterns being generated around me." It was, according to Gysin, the first work of art meant to be looked at with the eyes closed.

Sommerville's account is fascinating for several reasons, not least of which is its illustration of how the material culture of midcentury brain science, once it entered the public realm, could travel across a wide social landscape. Strobe became one of the best traveled features of that culture. As we have seen in earlier sections of this book, following Grey Walter's accounts of it in connection with his epilepsy research, versions of strobe found their way into contemporaneous accounts of madness and the early experiments of Scientology (see Clinical Tales 3 and 6). Sommerville's Dream Machine is yet another of the variations, and it in turn has parallels to the stroboscopic lamp described by the Hungarian dissident Lajos Ruff as part of the paraphernalia of the so-called magic room in which he underwent interrogation (see Clinical Tale 5). Those connections serve as further reminders of the crossover between brainwashing and the counterculture's fascination with hallucinatory experience.[*]

It was among Burroughs and his friends that flicker found its most devoted adherents. Their fascination with its vision-inducing properties is of a piece with their concern about its use by the "police." As Burroughs subsequently related in The Job (much of its text dating from the early

---

[*] Burroughs's writings on strobe echo an early piece written by the Situationist Guy Debord that cites Hebb's research on sensory deprivation and Ruff's book The Brain-Washing Machine as examples of the dark possibilities created by new "means of conditioning" and concludes by calling for artists to seize control of those means from the "police." "The Struggle for the Control of the New Techniques of Conditioning," translated by Reuben Keehan, Internationale Situationniste, June 1958. (See Clinical Tale 5.)

to mid-1960s, although not published until 1974), subjects exposed to flicker reported "dazzling lights of unearthly brilliance and color. . . . When the flicker is in phase with the subject's alpha rhythms he sees extending areas of colored pattern which develop throughout the entire visual field, 360 degrees of hallucinatory vision in which constellations of images appear." Here as elsewhere in Burroughs's work, the possibilities of visionary experience existed in close proximity to other possibilities that served the contrary interests of authoritarian "control": "Rhythmic sound, films and TV impose external rhythms on the mind, altering the brain waves which have otherwise been as individual as our fingerprints. . . . It is entirely possible that the EEG records of a generation of TV watchers will be similar, even identical."

What was the source of Burroughs's fascination with flicker? To answer this question, it is helpful to step back and recall some aspects of the story traced in previous chapters, particularly those relating to the topic of brainwashing, a topic of great interest to Burroughs. As we have seen, influential midcentury accounts of mind control argued that it was based on techniques that the Soviets had adapted from Ivan Pavlov to their campaign of mass ideological conversion. But as the 1950s unfolded, accounts of the transformations under way in postwar American society often incorporated a similar analysis, stressing the use of such techniques in the field of public relations that had become such a significant force within midcentury capitalism. By the time of *The Manchurian Candidate* the distinction between the two accounts had all but collapsed. "Conditioning," the methodical, purposeful process of molding human brain patterns, thought, loyalty, and behavior, described a phenomenon that many felt had become part of the lived experience of modernity on both sides of the Iron Curtain.

The Beat movement to which Burroughs belonged made this phenomenon central to its literary experiments. As the novelist John Clellon Holmes wrote in 1958, the movement belonged to a generation marked by a unique historical constellation: "This group includes veterans of three distinct kinds of modern war: a hot war, a cold war, and a war that was stubbornly not called a war at all, but a police action [i.e., the Korean War]." It was the first generation in US history, he wrote, that has grown up with peacetime military training as a fully accepted fact of life. It is the first generation for whom the catch phrases of psychiatry have become such intellectual pabulum that it can dare to think they may not be the final yardstick of the human soul. It is the first generation for whom genocide, brain-washing, cybernetics, motivational research—and the resultant limitation of human volition which is inherent in them—have been as familiar as its own face. It is also the first generation that has grown up since the nuclear destruction of the world has become the final answer to all questions.

With admirable concision, this passage sketches the midcentury constellation of forces that shaped the culture of the 1950s and the many mutant offspring they produced, among them the sixties counterculture.

It is within that context that Burroughs's obsessive quest for techniques of "deconditioning" belongs. His interest is manifest throughout his writings, starting with his very first efforts. *Queer* (written in 1953 though not published until 1985) concludes with a trip to South America in search of yage, a mysterious drug to which Burroughs attributed both visionary and mind control properties. It was Burroughs's belief that during the early years of the Cold War, both the Russians and the Americans had experimented with yage to determine its potential as an interrogation tool. In a letter to Ginsberg written in 1956, Burroughs called "brainwashing, thought control, etc., the vilest form of crime against the person of another." Yet as many of the scientists of this era had begun to demonstrate, these practices—and the hallucinatory states they often produced, which mirrored those frequently experienced in conditions such as epilepsy, migraine, and schizophrenia—also opened portals to realms where thought seemed to escape normal control. If most brain activity had hitherto taken place largely offstage, the conjuncture wrought by midcentury brain science and the Cold War now brought it to center stage in ways that proved tantalizing to figures such as Burroughs.

His subsequent writings return repeatedly to that conjuncture. They illustrate the way that interrogation and the altered states it produced in the subject had become a primal scene not just of the Cold War sciences of brain and behavior but of the counterculture as well.* The hallucinatory sequences that make up much of *Naked Lunch* include passages that update the dystopian science fiction scenarios of Huxley and Orwell by means of the radical new methods developed in the 1950s. In a scene that echoes the opening of *Brave New World*, Dr. Benway, the mad scientist who is introduced to the reader as an "expert in interrogation, brainwashing, and mind control," guides his visitor around his Reconditioning Center and discusses his techniques for manipulating patients via drugs, hypnosis, electricity, and so forth. A later passage raises the possibility of extending EEG research into "biocontrol": "control of physical movement, mental processes, emotional reactions and *apparent* sensory impressions by means of bioelectric signals injected into the nervous system of the subject." Here Burroughs offered the reader a classic variant on the midcentury paranoid style, evoking in satirical form the vast "machinery of influence" (as Hofstadter put it) that haunted the midcentury imagination.

But paranoia was also generative. Nightmarish as they were, such scenarios formed the backdrop to Burroughs's quest for a means of achieving what he called "freedom from past conditioning." He pursued that quest along multiple pathways. One of them, as we have seen, lay through Scientology. *The Job* contains an extensive discussion of Hubbard's methods of rendering its followers "clear" following "erasure" of their so-called Reactive Mind. The contents of the Reactive Mind (R.M.), wrote Burroughs, are made up of propositions that have "command value" at the automatic level of behavior. He likened them to the metabolic commands that regulate breathing, body temperature, brain waves, and so on and associated the regulatory center with the hypothalamus: "the neurological intersection point where the R.M. is implanted." What was the mechanism of that form of conditioning? Earlier civilizations such as the

---

* Chris Marker's film *La Jetée* manifests the same fascination with the connections between interrogation, memory, and hallucination. Partly inspired by accounts of torture during the war between France and Algeria, the film was released in 1962.

Mayans, he wrote, had achieved it through their minutely regulated and rigidly enforced calendars; nowadays, it was largely the function of the constant messaging of the modern mass media, which through a barrage of images and words cast a spell over audiences so powerful that they were left as if "imprisoned." He compared the "silence" that followed Hubbard's techniques of erasure with the perceptual isolation achieved in "sense-withdrawal chambers"—a condition that was terrifying to many but welcome to others such as himself: "Personally . . . it can't get too quiet for me."

Another pathway lay through the set of practices he became acquainted with through his reading of Grey Walter. In Burroughs's writings of the early 1960s, flicker and the effects induced by it are a constant presence (Dr. Benway's face even "flickers like a picture moving in and out of focus"). In his laboratory, as we have seen, Walter had been able to experimentalize epilepsy, to make this condition, whose uncanniness lay in the complete loss of bodily control brought on by a seizure, something that could be reproduced and studied with the aid of the stroboscope and EEG. But Walter had also come to believe that the epileptic phenomena experienced as flicker effects—whirling spirals, hallucinations, dizziness, memories, and disturbances of the time sense—could, with practice, be modified and controlled by the subject: "The will of the subject can be brought into play; he can, for instance, consciously and with effect resist or give way to the emotions or hallucinations engendered by flicker, a matter of no little social interest as well as enlightenment on the question of self-discipline."

It was this possibility that most fascinated Burroughs. In *The Job* he envisioned carrying it out on a grand scale and repeatedly expressed interest in the results of recent experiments in "mass deconditioning." One set of experiments that updated Pavlov's studies on dogs suggested that subjects could acquire conscious control of the processes normally regulated by the autonomic nervous system (blood pressure, breathing, and so forth). In another such trial, subjects wired to an EEG unit could, with practice in controlling their brain processes, learn to produce at will the alpha waves associated with relaxation. Burroughs's turn to such possibilities built on Walter's recognition of the adaptability and plasticity of brain waves to propose far-reaching changes in conscious-ness and identity.

Burroughs's long-standing interest in medicine was reflected in a life-time of reading medical literature on topics such as addiction, epilepsy, schizophrenia, and hallucination. During the late 1960s, he immersed himself in the new field of biofeedback, which practitioners had devel-oped as a tool for treating syndromes such as stress, migraine, even (some claimed) epilepsy. The method's origins lay in Hans Berger's discovery of EEG and Grey Walter's subsequent refinements of the technique. Electronic signals displayed on a screen allowed the subject to observe his or her brain-wave activity as though on a television set, and, over time, learn to regulate parts of the EEG spectrum. Brain-wave training turned what had been thought of as involuntary activity into the object of con-scious control. Burroughs saw it and related methods as offering a means of regulating the visceral processes of the body's internal environment, including brain waves and other automatic functions (breathing, internal temperature, and so on) that normally escaped conscious awareness. It was, so to speak, a physiological or metabolic unconscious, quite different from the Freudian unconscious. Burroughs carried out experiments with a device called an "Alpha Sensor" and reported interesting effects in con-nection with the use of a feedback system for synchronizing flicker with alpha rhythms.* Scattered references in his letters and notes also seem to indicate that he saw this method as a kind of hybrid of Hubbard's and Walter's techniques. He described the E-meter as a biofeedback device that, in combination with the repetitive commands that accompanied auditing, functioned much like a form of electric brain stimulation, one that could even produce "pictures and films." Left unanswered were a series of questions: What was the source of the images? Were they memories? dreams? visions of something else entirely?

---

* "Biofeedback tells us more about what is going on inside our own bodies. It is a 20th century extension of that time honored dictate 'Know Thyself.' With biofeedback instru-ments you can not only know something about the electrical activities within your body but learn to control them in a manner hitherto thought impossible for all but a few Indian Yogis." "Aleph One" pamphlet, William S. Burroughs Papers, New York Public Library.

Burroughs's interest in such techniques stands in sharp contrast to the many passages in his writings in which he sounded antipsychiatric themes. Citing Scientology's critique of modern psychiatry, he wrote:

> You don't have to tell me. . . . I know about the use of psychiatry committed as a weapon to eliminate anyone the establishment wants to get rid of. We have seen this in Russia and Germany. I am violently opposed to shock treatment, lobotomy, and a new form of cerebral castration by burning out the sex centers in the back brain. Most so called psychiatric institutions are simply death camps with not much pretense of being anything else.

Such echoes were far from incidental, for there was a good deal of shared ground here. At a major gathering of antipsychiatrists held in London in 1967 (the so-called Dialectics of Liberation congress) that brought together many of the luminaries of the movement, including Gregory Bateson and R. D. Laing, Allen Ginsberg took the stage to proclaim the need to fundamentally rethink the project of being human. He identified his friend Burroughs as a figure who "provides counterbrainwash techniques and leads the reader to examine conditioned identity." Deconditioning was a major theme of the event. Yet there were limits to the views Burroughs shared with the antipsychiatrists. As one scholar has observed, Bateson and Laing were largely uninterested in the biological brain; in denouncing the reduction of mind to brain, they wound up with a model of mind that was all but "brainless."

Burroughs, however, followed contemporary developments in brain science closely. The apparent reductivism of the new brain paradigm was deceptive, for the findings associated with that paradigm opened up pathways that he was only too eager to explore. He was deeply fascinated by states of madness such as schizophrenia and in that regard belonged, along with the antipsychiatrists, to a cultural tradition in which the mad were seen as possessors of an authentic—in Burroughs's term "nonconditioned"—vision. Yet his lifelong experience of addiction made him far more deeply attuned to the physiological dimensions of such phenomena. Here, too, in all likelihood, lay one of the sources of his fascination with flicker and its link to epilepsy. The hallucinations that accompanied the auras of epilepsy belonged to a spectrum of altered states that held immense interest for him.

This is illustrated in a short piece he published in June 1968 in the British men's magazine *Mayfair*, titled "Switch On and Be Your Own Hero." The piece was a satirical fantasy exploring the possibility that Walter's flicker technique could be adapted for use on a mass basis as a means of "identity switching." Images projected onto a person's face would enable him or her to assume the identity of anyone he or she chose: "your favourite pop star . . . your favourite actor living or dead." In a variation of the technique, the projection of X-rated movies onto the body would allow a person to act out pornographic scenarios. That in turn would provoke the inevitable response from the forces of control: Burroughs darkly imagined the Board of Health issuing an advisory that "The projection of sex films on flesh can give rise to venereal diseases hitherto unknown." As he explained in closing, the kind of mass hallucination via projector or magic lantern he described had been first introduced to the public by Brion Gysin and Ian Sommerville at a performance in Paris in January 1962 at which the Dream Machine had been unveiled. Its possibilities remained yet to be fully explored. In Burroughs's fevered fantasy, epilepsy served as a model for experiments in producing "nonconditioned" perception.

In harnessing the possibilities of flicker, Burroughs's friends Brion Gysin and Ian Sommerville believed that they had devised an invention of great historical significance. They envisioned the mass production of the Dream Machine and its eventual replacement of television, whose spot in a typical household it would assume. Gysin made especially large claims for the device. He believed that it would usher in an entirely new era of exploration devoted to the mapping of "interior space." An unpublished piece he coauthored with Burroughs, "The Golden Dreamachine," provides a glimpse of the extravagant hopes they pinned on it. It tells the story of Johnny Space, a boy who has grown weary of television and prefers the dreams in his head. Johnny finds instructions for a Dream

Machine in the attic of his home; once assembled and switched on, the machine "bursts like a kaleidoscope in his head." He sees a small man who identifies himself as the Technician, come to take Johnny to space: "I will show you the ship which you have inside your head already." The figure reappears in another short piece written by Burroughs, "Operation Sense Withdrawal," that includes fragmentary images of immersion tanks and "vast flicker cylinders and projectors" sweeping a city landscape with light.

Inspired to speculate along ambitious lines, Gysin linked flicker to a further stage in an evolutionary process culminating in a completely transformed state of consciousness. Once again the altered states associated with this new form of consciousness were connected to the mass media. Interior space, it appeared, was composed of inner movies. Gysin likened the dreamlike visions in the alpha band that were induced by flicker to motion pictures and mused about their source: "Who can say who is projecting these films? Where do these films come from?" As though channeling the neurosurgeon Wilder Penfield, he proposed one answer: "One gets flashes of memory, one gets these little films that are apparently being projected into one's head," adding "one then gets into an area where all vision is as in a complete circle of 360 degrees, and one is plunged into a dream situation that's occurring all around one."

As we'll see in the final chapter, which traces John Lilly's experiments with sensory deprivation, other figures of the emerging counterculture posed versions of the same questions asked by Gysin. Their answers varied widely. Marshall McLuhan, in his efforts to undo the conditioning effects imprinted by the modern mass media, sought to help the public to "observe consciously the drama which is intended to operate on it unconsciously." Here he was referring to a primarily externally generated drama. The counterculture radicalized that by cultivating methods designed to make conscious the *internal* drama that normally operated at a subconscious level: to attend, in effect, to the "internal movie." The way had been led, as we saw in earlier chapters, by Wilder Penfield and Grey Walter in their work with patients with epilepsy. Both figures drew on the metaphorical possibilities of the mass media to capture the myriad forms of complex internal imagery elicited by their techniques. What took shape in the 1960s was a deliberate effort to activate and observe such imagery, an effort that repeatedly came up against Gysin's questions: Who can say who is projecting these films? Where do these films come from?*

Gysin was briefly able to attract the interest of prominent figures in the media industry, but efforts to find committed backers for his plan to go mainstream with the Dream Machine ultimately failed. Yet as flicker, in the form of neon, fluorescent lighting, and television, became an ever more pervasive feature of the everyday media and social environment, artists and cultural entrepreneurs followed Gysin and Sommerville's lead in trying to harness it more purposefully. At concerts, in dance venues, and in other settings, strobe became part of the visual surround of the sixties. Further offshoots of the flicker aesthetic found their way into the world of experimental filmmaking. In 1965, the artist and composer Tony Conrad, who had absorbed the work of Grey Walter while a student at Harvard, released a short film, "The Flicker," which sought to generate effects similar to those produced by the Dream Machine. The film consisted simply of two frames, one black and one white, that alternated at varying rates to produce stroboscopic effects. It was prefaced by a disclaimer of responsibility for any "physical or mental injury" it might cause and cautioned audience members that they remained in the theater at their own risk because the film might trigger an epileptic seizure. Through such means a generation of innovative filmmakers invited audience members to undergo experiments similar to those conducted by Grey Walter in his Bristol clinic.

---

* In Stanisław Lem's novel *Solaris*, scientists are sent to observe a planet whose surface is covered by an ocean or plasma that is likened to a brain. In the conditions of sensory deprivation that mark life on their spaceship, they find themselves experiencing vivid hallucinations whose source is enigmatic. Are they symptoms of mental illness? Memories? Emanations from the planet itself? Here, too, epilepsy is invoked in relation to the mysterious properties of the ocean brain. Lem, *Solaris*, translated by Joanna Kilmartin and Steve Cox (New York: Walker, 1970 [1961]), 174–75.

CHAPTER 7

Brain Science Between Cold War and Counterculture

The Case of John Lilly

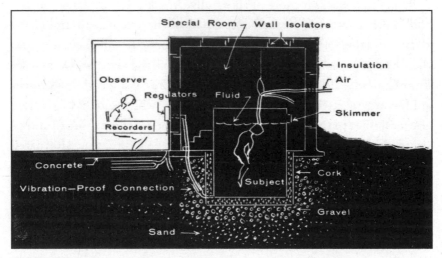

Isolation tank designed by John Lilly and Jay Shurley.

Lewis Carroll, the migraine-afflicted creator of Alice, the mushroom-eating little girl of his Wonderland stories, was a kind of patron saint of mid-twentieth-century brain science. He was frequently invoked by figures such as Grey Walter, Norbert Wiener, and Donald Hebb, who seemed to see him as a Victorian-era explorer of the inner world whose map they did so much to fill in in the 1950s. Another of those cartographers was the Russian neuropsychologist A. R. Luria, who described that fantastical world in connection with his patient S., the subject of *The*

Mind of a Mnemonist: A Little Book About a Vast Memory (1967). S.'s astounding powers of recall and gifts of inner visualization continually evoked for Luria "the feeling little Alice had after she slipped through the looking glass and found herself in a strange wonderland."

Though visions of this neurological wonderland tantalized many midcentury scientists, few went farther in exploring it than did John Lilly, himself a lifelong *migraineur*. Now best known as a dolphin researcher, the author of countercultural classics such as *The Center of the Cyclone*, and an avatar of expanded consciousness, Lilly began his career as a central figure in post–World War II brain science. Perhaps no single person better embodies the continuities between that field, its offshoots in CIA-sponsored investigations into mind control (his contributions to which are cited in the agency's interrogation manual *KUBARK*), and the practices of self-experimentation and consciousness exploration that are one hallmark of the 1960s. If Grey Walter might be seen as the March Hare of midcentury brain science, John Lilly was its Mad Hatter. Both bequeathed significant legacies to the counterculture: flicker in the case of the former, the isolation tank in the case of the latter. Whether through the overstimulation of one sense or the understimulation of all senses, both opened doorways to visionary experience.[*]

Lilly began his scientific career positioned at the leading edge of midcentury neurophysiology, doing advanced work on electrical brain stimulation. He embraced an ambitiously multidisciplinary ethos from early on. One sign of this was the psychoanalytic training he underwent as part of his quest to find ways of bridging the divide between the sciences of brain and mind. That quest, which consumed him until the end of his life, took him down increasingly unconventional paths. Prior to that, however, he spent close to two decades occupying a position at the heart of the government–behavioral science nexus. Given the historical context, it was inevitable that his research, first on implanting

---

[*]   Both also saw themselves as breakers of taboos. As Lilly wrote, areas of ignorance about man in relation to sex, pain, and torture have "taboo" signs built around them prohibiting research in such areas. Lilly, "No Holds Barred," n.d., John C. Lilly Papers, 5A1, Box 5, Stanford Library. Like Walter, he sought to dismantle such prohibitions.

electrodes in the brain, then on sensory deprivation, attracted the interest of the US intelligence community. His mounting qualms about that led him, at the end of the decade, to turn his back on government employment and commit himself to increasingly far-out investigations involving isolation tanks, dolphins, and psychedelics. How can we make sense of his trajectory?

Hints of answers can be found in Lilly's early life, details of which are included in his autobiography *The Scientist* (1978). The account describes a precocious boy with a troubled personality. It is marked by lengthy episodes of illness, migraine, and dissociative spells, during which the imaginative features of his persona ran wild. His juvenile efforts at writing science fiction provided one early outlet for his febrile imagination. One of the major conflicts of his childhood arose from his struggles with the deeply Catholic family milieu in which he was raised. An early moment of rupture in his life came with his encounter, during his first year of classes at California Institute of Technology, with the news of the "death of God" and his subsequent immersion in the writings of Darwin and Freud, as well as Aldous Huxley's *Brave New World*. Yet even as he embarked upon his scientific studies, distancing himself increasingly from Church doctrine, he found nonworldly substitutes for God: "guardian angels" and "Beings" from other planets or dimensions. Those, as we shall see, remained lifelong companions.

Another major turning point came with his switch from physics to biology and then to neurophysiology. In that respect his trajectory mirrored the larger scientific trends of the era. As earlier noted, physics was seen as queen of the sciences for much of the twentieth century, a status confirmed by the dramatic success of the Manhattan Project. Yet mid-century developments laid the groundwork for the emergence of the life sciences as an increasingly significant field in their own right. The discovery of the helical structure of DNA in 1953 was one major milestone of that development. It was crowned by the awarding of the Nobel Prize in Physiology or Medicine to Francis Crick and James Watson in 1962. In a speech given that year, Crick predicted that the scientific research of the future would be dominated by the question of the relationship between brain and mind, in particular, the biological basis of consciousness.* Lilly's contributions to the emergent paradigm of the "explorable brain" were dedicated to answering that question.

Known to childhood friends as "Einstein, Jr.," Lilly made the switch from physics to biology, he later recounted, following his reading of a paper by the Cambridge neurophysiologist Edgar Adrian, Grey Walter's adviser, that confirmed the significance of Hans Berger's newly invented EEG. That provided the germ of what he eventually came to think of as his "mission": precisely the integration of brain and mind that Crick spoke of in 1962. Though he conceived of that mission early on, it was only after World War II that he found himself in a position to pursue it. During the war he worked on studies of high-altitude sickness commissioned by the air force. Those studies were the first of a long series of investigations into the functioning of the human organism in extreme states: POW camps, space travel, and isolation tanks. Those investigations were the corollary, it might be said, to the "age of extremes" inaugurated by world war, nuclear power, and superpower conflict.

Designing a "Brain-Television-Set"

Lilly began the 1950s straddling two different scientific communities. With a position at the National Institutes of Health, where he held dual appointments at the National Institute of Neurological Diseases and Blindness and the National Institute of Mental Health, his world was split, as he later put it, between two views of man as possessor of a brain and possessor of a mind. He spent much of that decade crossing back and forth over that divide in search of ways of bridging it and forging connections with like-minded figures. One key ally was a figure we have encountered before, the psychoanalyst Lawrence Kubie. Kubie nursed ambitions

---

*   One member of the audience at that speech was the young medical student Oliver Sacks, who later wrote that the scientific study of consciousness had hitherto been impeded by the belief that it was totally subjective and thus off limits to research. Sacks, *On the Move: A Life* (New York: Vintage Books, 2015), 334.

similar to Lilly's, and the two became close friends. One expression of
their closeness is to be found in Lilly's highly appreciative comments on
Kubie's paper about the psychoanalytic word association experiments he
had conducted with Wilder Penfield's epileptic patients while they underwent surgery (see chapter 2). Citing his own first encounters with the
work of Adrian and Freud as having sparked his interest in integrating
neurophysiology and psychoanalysis, Lilly went on to voice high praise
for Kubie's contributions to that project of integration. Those built on
Penfield's finding that electrical stimulation of the temporal lobe could
evoke the "reliving and re-observing of past events." Whether they were
repressed memories remained to be established, as did their relation to
underlying neurotic processes. Lilly called for further research on these
matters and noted his own preliminary efforts. Whereas Penfield used
only a single electrode in his neurosurgical operations on patients with
epilepsy, Lilly used a multitude of electrodes in experiments with mon-
keys whose results were then captured for analysis by an apparatus he
referred to as a "brain-television-set." He also made passing reference to
some of the early results of his experiments, which included what he cryp-
tically referred to as "apparitions."

The early 1950s were a period of intense ferment for Lilly, in which
he continually weighed and discarded options for recording the brain's
electrical activity. In a letter written in the fall of 1951 to Robert Morison
of the Rockefeller Foundation, to which Lilly had applied for funding,
he outlined a research program predicated on the search for neurophysiological correlates for all psychological activity. Morison responded by
pointing out to Lilly the difficulties facing him and reminded him of
the eminent British physiologist Charles Sherrington's methodological
caution that brain and mind were two quite different things. But he
ended on a more encouraging note: "You are one of five or at most ten
people alive today who have enough knowledge of neurophysiology and
psychiatry to lead us out of the wilderness."

How was Lilly to advance this project of integration? The first
fruits of his efforts included the invention of the aforementioned TV-
like apparatus for recording the brain's electrical activity. Unlike Grey
Walter's toposcope, which obtained readings produced from the skull's

surface, Lilly inserted electrodes into the brain itself, adapting the corti-cographic techniques pioneered by Wilder Penfield and Herbert Jasper. The results were then displayed on a screen. He described the machine to Morison as a twenty-five-channel amplifier that, by registering elec-trical activity from twenty-five positions, produced "moving pictures" of brain activity. That was already a significant accomplishment.* Yet con-fronted with the technical difficulties of establishing enough channels to cover all brain regions, Lilly entertained far more radical possibilities. He envisioned implanting hundreds, even thousands of electrodes into a living brain. Even that represented only one facet of what he conceived of as an undertaking that would encompass a huge array of scientific disciplines. His conviction that contemporary psychiatry had little to offer when it came to knowledge of the mind led him to become a mem-ber of the Philadelphia Association for Psychoanalysis and to undergo his own analysis. Any remaining gaps in correlation between brain and mind, he wrote Morison, would best be addressed in a neurosurgical setting such as Penfield's Montreal Neurological Institute. Along with the essential disciplines of neurophysiology, neurosurgery, and psycho-analysis, he listed cybernetics, engineering, and mathematics as key auxiliary fields. Disdaining methodological caution, he envisioned a kind of unified field theory rivaling Einstein's. He wrote to Morison that the brain didn't recognize disciplinary boundaries but simply "goes on its merry way."

Some of the flavor of what he envisioned during that period, with-out ever carrying it out, is captured in *The Scientist*, which recorded the following exchange with his psychoanalyst (Lilly's adoption of a third-person narrative for his autobiography, at first glance strange, is perhaps best understood as a strategy for preserving scientific objectivity vis-à-vis the radically subjective content of much of the narrative):

---

* A later version of the "brain-television-set" that used noninjurious current was her-alded by NIMH director Robert Felix as the "single most exciting discovery in the field of brain physiology." See Lilly biography, *Who's Who* (London: Adam and Charles Black, 1962), Box 29 John. C Lilly Papers.

ROBERT [WAELDER, PSYCHOANALYST]: So you want to hook your
brain up to a recording system and then play that recording back into
your own brain. Is that correct?

JOHN: Yes. I visualize ten thousand, a hundred thousand, a million
electrodes inserted through my skull into my own brain, hooked up
to an adequate recording system. The records would be made while
I am actively doing something. At a later time they would be played
back through the same electrodes. I would then be able to tell if I went
through the same actions, also recorded on motion picture film, that
I had gone through originally. I would also be able to see whether or
not I had the same inner experience that I had during the original
recording.

Implicit in this statement was an ethical imperative. Lilly described
himself as having been trained in the school of human physiology in which
the scientist himself is the first subject of the experiment and added that
until one was willing to undergo the experiment oneself, one should not
perform it on other humans: "A doctor should not insert brain electrodes
in patients unless he is willing to insert them in his own brain." Apart
from the ethical problems raised by such a procedure, he feared that the
results would be compromised by damage to the organ. A brain with ten
thousand electrodes in it would necessarily be an "injured" brain.

The ethical disquiet surrounding Lilly's research was greatly ampli-
fied by the Cold War circumstances in which he came of age as a scientist.
Much of his work developed within the matrix of what Alfred McCoy
called the "Manhattan Project of the mind" launched in the early years
of the Cold War, in which the potential for weaponizing many branches
of clinical and behavioral science was explored. Once he began to report
on the results of his electrode research, it was not long before the US intelligence and military communities began inquiring about its potential
operational uses. In a chapter of his autobiography titled "Control of the
Brain and the Covert Intelligence Services," Lilly told the story of his
encounters with various figures from those branches of government and
the doubts he had begun to experience as he found himself combating
pressure to participate in classified briefings. He sought to resist such

pressures but found himself continually outmaneuvered by representatives of secretive government agencies whose mandate, it was clear, was to gather data about all methods that might confer advantage in the age of "brain warfare." Lilly's highly advanced work on such areas as the pain and pleasure circuits in the CNS promised just such an advantage. With due allowance for the occasionally stagy way in which he told the story, its outlines made clear the ethical hazards created by the possibilities for funding and recognition that the Cold War placed in front of people of Lilly's abilities.

Whether as the result of his qualms about the potential for abuse of his electrode research, his realization that the scientific value of results obtained by turning the brain into a sort of pincushion was questionable, or simply his desire to continually extend experimental boundaries, he then moved into the field of research for which he became best known: sensory deprivation. The Rockefeller Foundation's Robert S. Morison took credit for this move, noting in a diary entry in 1954, "John Lilly has been sufficiently excited by his visit to Donald Hebb (on Morison's recommendation) that he has undertaken to repeat the human isolation experiments." Lilly remained in touch with Hebb throughout the 1950s. However, he adopted a different technique in order to minimize the effects on proprioception of forces such as gravity, which were not taken into account in Hebb's setup. Whereas Hebb sought less total sensory deprivation than depatterning and monotony, Lilly sought to isolate the subject as completely as possible from all incoming stimuli by immersing him in a tank of body-temperature salt water housed in a room that was pitch dark and dead silent. He claimed that within "two or three hours he can get almost all the effects that Hebb has reported in 48 hours." Lilly spoke from personal experience, acting as the subject in these experiments and reporting vivid hallucinations and many other "bizarre psychological changes." It was his conviction, as he told Morison, that the fact that he had undergone psychoanalysis allowed him to face those changes without undue anxiety.

Sensory deprivation became, it seems, Lilly's solution to many of the problems that had hitherto plagued his research. At a symposium held in 1955, he expressed his view that isolation tank experiments would become one of the cornerstones of a genuinely multidisciplinary research program combining, at a minimum, the approaches of neurophysiology and psychoanalysis. Reviewing the results of Hebb's experiments, he noted the finding that subjects in isolation experienced heightened suggestibility and hallucinations similar to dreams or the visions associated with mescaline. He speculated that the desire for stimuli experienced by them resulted in the projection of internal visual imagery and added that the results potentially had implications for inquiry into both psychotherapy and brainwashing. Among the important conclusions he drew from his isolation tank experiments was one central to the emerging picture of continuous brain activity. At the beginning of the 1950s, some scientists in the field had posited that the brain's natural state in the absence of external stimuli was sleep or coma. Lilly's solitude experiments and the visionary experiences they produced provided powerful evidence that consciousness was not dependent on external stimuli but lay within the "natural rhythms of the brain's cell circuitry itself."

As the passing reference to brainwashing in the above comments attests, the pure state in which Lilly sought to isolate the brain was inextricably bound up with the geopolitical dramas of that time. It was likely as a result of his expertise on the topic of isolation, which had become central to investigations into brainwashing, that in 1956 he was asked by the Group for the Advancement of Psychiatry (GAP) to organize two symposia on the topic of forced indoctrination. The programs of the symposia, both held in Asbury Park, New Jersey, featured several figures who have appeared previously in these pages, including Harold Wolff, Lawrence Hinkle, Jolyon West, Robert Lifton, and others. They met in a climate in which the hysteria surrounding brainwashing had once again become full blown. This was the year Edward Hunter's second book (*Brainwashing: The Story of Men Who Defied It*) appeared in print, and its impact on the national imagination was clearly on Lilly's mind as he opened the proceedings of the second symposium in the fall of that year by invoking George Orwell's *1984*:

> This morning we are considering the psychological processes of thought reform and indoctrination with force, or "brainwashing,"

or whatever you wish to call these processes. The term "brainwash-ing" has come apparently to have a meaning created by the national press. In this meaning the term is more or less defined by an indoc-trinator in the following quotation [from *1984*]; the indoctrinator is talking to his victim: "We make the brain perfect before we blow it up." ...

George Orwell not only wrote a great novel, *Nineteen Eighty-Four*, but he wrote a disturbing handbook for brainwashers which convinced his readers of the reality of an extreme form of forcible indoctrination.

After enumerating the procedures that Orwell's protagonist, Winston Smith, is subjected to—isolation, starvation, torture, drugs, direct electrical stimulation of his brain—Lilly observed that his por-trayal of the "science" of brainwashing had been adopted as a kind of definitive picture of what had been done to US servicemen in North Korean POW camps. It was therefore vitally important for the presenters "to clarify the differences between the fantastic account of Orwell and the real processes actually used in authentic cases." Most of the participants were in agreement, having read the lengthy paper by Harold Wolff and Lawrence Hinkle (both present) that stripped away the scientific veneer of brainwashing (see chapter 4). The psychiatrist Robert Lifton, for in-stance, elaborated on the concept of "milieu control," or the manipulation of communication in the POW camp environment: *1984*, he observed, presented a version that stressed mechanical means (a two-way telescreen enabling the government to surveil its citizens), whereas the Chinese used a purely "human recording and transmitting apparatus" for producing confessions and molding thought. Later in the proceedings, Lilly alluded once more to the power of the cultural fantasy surrounding mind control, observing, "I believe that the literature that is coming out at the present time, including that which we have heard here today, can be said to be a handbook on what brainwashing is rather than on what one fantasies it can be."

Yet at least some of the participants continued to push for clearer answers. The specter of mad science that was so central to the public image of brainwashing proved difficult to banish. In response to one participant's question about the Pavlovian basis of Communist interrogation, Lifton pronounced it to be a myth that had originated in journalistic circles, clearly referring to Edward Hunter. Other questions, however, were harder to answer, going directly to the heart of Lilly's own research. One air force physician began by acknowledging Wolff and Hinkle's argument that scientific methods had not so far been used by the Eastern Bloc but then switched to a more Huxleyan vein, stressing the need to recognize "a potential future which may be decades off but which nevertheless is foreshadowed by developments like the work of Dr. Lilly and Dr. Hebb and others on sensory deprivation, work on psycho-pharmaceuticals . . . and work on electrical stimulation of the brain."

According to his autobiographical account, Lilly became increasingly haunted by those future possibilities, and within a few years he severed his ties with the government. Even as he did so, however, the notion of brainwashing assumed a new significance within his own thinking. In an unpublished paper from the mid-1950s, he seemed to side with the dystopian vision of Aldous Huxley over that of Orwell.* In considering the possibility of chemical and electrical thought control, he expressed the view that *Brave New World* "will not be imposed on us by Big Brother. We will beg for the raptures of Soma and call it a sacrament." If it were ever to be fully realized, in other words, mind control would not be coercive but voluntary: as Huxley had warned, people would come to love their subjugation. Though in his public pronouncements Lilly treated the term *brainwashing* as ill chosen and cast the premise of *1984* as science fiction, privately he himself proved far from immune to scenarios such as those depicted by Orwell. As we shall see, his own increasingly speculative

---

* See Lilly's letter to Aldous Huxley's wife, Laura, of September 17, 1968, spelling out his intellectual debts to her husband, who became a personal friend and an ally of his dolphin project. Huxley (Aldous and Laura) Papers, box 26, folder 14, UCLA. (My thanks to Jeffrey Matthias for sharing this letter with me.) Fascinated by dolphins' intelligence, Lilly devoted much of his later career to trying to communicate with them. See D. Graham Burnett, "A Mind in the Water," *Orion*, May–June 2010, 38–51.

thinking relied on a model of thought control in order to conceptualize what he came to call "human consensus reality."

Lilly in Wonderland

It is in connection with his research into sensory deprivation that Lilly's work took the turn that exemplifies John Marks's observation that mid-century research on brain and behavior had both a mind control side and a transcendental side. Lilly became a foremost specialist in the field, and his work would be cited in *KUBARK*; he also played a principal role in introducing sensory deprivation to the sixties counterculture. At a conference on sensory deprivation held at Harvard Medical School in 1958 that attracted figures such as Hebb, Kubie, Walter, Wiener, West, and others, Lilly chaired a session. In a sign of his growing alienation from the scientific mainstream, however, his presentation was not included in the published proceedings and his remarks in the discussions were edited out. As he explained in a letter to Kubie, his strenuous objections to the use of the term *sensory deprivation* succeeded in alienating the conference organizers. His preference for the term *isolation* lay in the fact that, as he explained, "I myself experience very little, if any, 'deprivation' in the water tank experiments."*

Beginning with Hebb's original investigations, the experimental setup underwent a rapid evolution from chamber to box to tank. Whereas Hebb emphasized the depatterning of external stimulation, or monotony, other researchers, including Lilly, sought to reduce external stimulation to an absolute minimum. Lilly and his colleague Jay Shurley conceived of their work as an "experimental intervention into the complex feedback loops normally existing between individual and environment." Many researchers, though not all, found that their subjects reported having experienced hallucinations and that they generally

---

* Himself increasingly mystified by the path Lilly was following, Hebb wrote to him in 1959 to ask "What are you up to?" Hebb to Lilly, March 12, 1959, Hebb Fonds, 2364.01.20.2-0037.

followed a pattern ranging from simple to complex. Parallels were found not only in accounts of the extreme monotony and disorientation experienced by sailors, pilots, and prisoners but also in clinical accounts of epilepsy, migraine, drugs, hypnosis, and stroboscopic flicker. According to Hebb's student Woodburn Heron, hallucinations, defined as "perception without object," experienced in sensory deprivation shared many characteristics with drug-induced hallucinations as well as those produced by Grey Walter by means of flickering light. In his presentation at the Harvard symposium, Lawrence Kubie treated sensory deprivation from a psychoanalytic perspective, noting the regressive dimension of "Lilly's experimental amniotic fluid" and (echoing his speculations concerning the hallucinatory reminiscences evoked by Penfield) suggesting that the experience provided access to memories buried in the unconscious.

To follow John Lilly along his subsequent scientific path is to venture, like Alice, through a veritable looking glass. It is a path that most scholars, mindful of his later reputation as a "piper of whale-hugging and cosmonaut of heightened consciousness," have been unwilling to follow him along, and given the pronounced strangeness of his later writings, that is hardly surprising. This also no doubt accounts for his virtually complete—and seemingly rather unjust—absence from the history of brain science.* What follows makes no claim to a complete exposition but confines itself to a few key points. Lilly's turn to isolation tank research seems to have reflected a confluence of professional, political, and personal factors. Among other things, it was a continuation of his quest to find ways of bridging the brain-mind divide without risking physical injury to the brain. In keeping with his friend Kubie's suggestions, Lilly evidently also understood his isolation tank experiments as a self-guided continuation of his psychoanalysis. In conditions of reduced sensory input, his mind was flooded by memory. Yet although the visual imagery he experienced in the tank included material drawn from his past, it also

---

* Lilly's work, for instance on brain stimulation, survives only in the form of footnotes in books such as José M. R. Delgado's *Physical Control of the Mind: Toward a Psychocivilized Society* (New York: Harper & Row Publishers, 1969). In this form it likely served as an unacknowledged source for the "brain-in-a-vat" thought experiment.

included material of decidedly more enigmatic origins. With its distinctive understanding of the unconscious, psychoanalysis provided only limited help in navigating that terrain. Other frameworks were required to help do so.

In "Experiments in Solitude, in Maximum Achievable Physical Isolation with Water Suspension, of Intact Healthy Persons," the paper coauthored with his colleague Jay Shurley that he gave at the SD symposium in 1958, Lilly addressed the central epistemological concern of his experiments. His research, as he put it in a telling phrase, was animated by the need to establish methods of observation capable of obtaining the "maximum of information from the subjective sphere." This quest for "new information" raised a number of important ethical considerations, insofar as it encroached on areas ordinarily held to be off limits. Elsewhere he described those areas (which included such topics as sexuality, torture, and pain) as "taboo, sacred, inviolable, by definition not to be investigated." In the absence of knowledge such areas remained enveloped in fear, superstition, science fiction, "fantastical zones of thinking." As a result of research inspired by Pavlov, he conjectured, the Russians might already have surpassed the West in that field.

Among the taboos he described was the behaviorist ban on introspection and the scientific study of consciousness. Lilly's experiments in solitude constituted a radical attempt to extend the boundaries of scientific inquiry into that realm. He went on to explain his conviction that the range of phenomena available to the normal human mind is "much greater than 'society' will apparently admit or accept." Therefore in his experiments the subject was given permission to experience whatever he could: "If you will, the individual is 'indoctrinated' in the principle of freedom for his own ego within the internal experiential sphere." Within the confines of the tank, the taboo against peering into the black box of consciousness could be lifted. For Lilly, self-experimentation was thus not only an ethical first principle but, methodologically speaking, the starting point of all scientific study of consciousness.

In elevating the quest for new information to a methodological absolute, Lilly also signaled his allegiance to a cybernetic model of brain and mind. The language of programming and reprogramming became

increasingly central to his work. Even as it did so, psychoanalysis continued to provide an important conceptual anchor. As he put it in a late letter
to Kubie, "I deeply feel that finally things are opening up and that we can
marry 'the objective' and 'the subjective' more effectively than we have
been able to do in the past." In a reference to the experiments Kubie had
carried out with Penfield in the early 1950s, he went on, "Your dream of
tying the kind of events seen in neurology and neurophysiology to mental
CS UCS [conscious and unconscious] events can be approached and possibly investigated in an effective, unequivocal and dynamic way." There
he echoed his comments on Kubie's paper of the early 1950s, in which he
had stated that Penfield's patients were indeed "reliving and reobserving"
past events, and reiterated his view that that work promised a genuine
synthesis of psychoanalytic and neurophysiological approaches. Concepts
of feedback, information, and programming became increasingly pervasive in his later work. Eventually, following his adoption of the model of
the human organism as a "biocomputer," he became convinced that the
mind-body dichotomy had been overcome. (By contrast, Penfield, who
also adopted such language, remained a dualist, claiming that the mind
acted independently of the brain in the same way that the programmer
did of his computer.)

The ethical constraints underpinning Lilly's isolation experiments
were greatly tested by the anxieties and pressures of the early Cold War
era and the resultant interest in knowledge of the factors that render individuals susceptible to forced indoctrination. Though Lilly regarded public fears on this score as exaggerated—telling Hebb in one exchange of
letters that he found press accounts of brainwashing to be "nonsense"—he
nevertheless took potential future developments very seriously. Hebb's
response to Lilly was to argue for ending the secrecy that surrounded
research in the area, a sentiment Lilly shared. Yet although his memoir
narrates a tale of growing disenchantment with government employment,
his personal papers tell a somewhat more ambiguous story about Lilly's
trajectory.

Consider, for instance, an unpublished document of his dating from
the mid- to late 1950s, with the title "Special Considerations of Modified
Human Agents as Reconnaissance and Intelligence Devices." The paper

conceives of human subjects as collectors, storers, and transmitters of information. In it Lilly notes that in conditions of isolation, internal generators of "signal data" (thoughts, feelings, memories, hallucinations) seem to be activated. Those data could be studied and analyzed. But further research along those lines, he wrote, was urgently needed to determine both the "new information" obtainable from *within* and the effect that the introduction of new information from *without* could have in conditions of isolation. That was especially true given what Lilly predicted would be the rapid acceleration of neurophysiological techniques for extracting information from and injecting new information into biological organisms "including man." In that context he made reference to Ewen Cameron's work on reprogramming (or "depatterning") his psychiatric patients. As we saw in chapter 4, Cameron had reported that in combination with LSD, electroshock, and sensory deprivation, "psychic implants" could be induced through verbal, emotionally laden tape-recorded statements repeated for days at a time and that they might exert profound effects on thought and behavior. Lilly evidently saw Cameron's therapeutic implants as simply one manifestation, produced under highly controlled circumstances, of a much wider process of programming and reprogramming whose agents included not only parents, church officials, and teachers but also more malign forces. He noted, for instance, the parallels of that therapeutic approach to police state interrogation methods in which isolation and verbal and written repetitions were part of the regimen. In addition to their uses in therapy and interrogation, such techniques "are used in the long-term indoctrination of populations in newly acquired countries under the Communists."

A subsequent section on electrical stimulation of the brain made many of the same points. He speculated that techniques would soon exist allowing thoughts to be controlled and then experienced "as if voluntarily self-controlled," before then going on to invoke an overtly science fiction scenario: "This could lead to master-slave controls directly of one brain over another." Investigations into that possibility were, he claimed, planned in his laboratory. In another unpublished text from around the same time titled "Human Brain Research and Ethics," he conjured up another version of that dark scenario and concluded by asking "Should a

Manhattan Project on mind control be set up by electrical control of the brain?" He warned of the need for appropriate preventive measures to avoid a future in which the visions of *Brave New World* and *1984* might be realized. Again, though the chronology remains unclear, these documents suggest a degree of ongoing engagement in the theory and practice of mind control difficult to square with the narrative presented in his autobiography.

It is striking that precisely at the point at which he confronted the most radical of the possibilities engendered by his own research, Lilly was simultaneously making the turn to the quite different possibilities conjured by his friend Huxley's writings on "Mind-at-Large," according to which the brain's constant drive for optimal efficiency can be disrupted by psychedelic drugs, thus opening it to perceptions ordinarily blocked from conscious awareness. "Is the brain," as he asked in his autobiography, ". . . a valve for a universal consciousness?" Information, understood in the enlarged sense he wanted to give it, could travel along many strange frequencies, ranging from direct brain-to-brain communication along a hundred channels to communication with other beings, including dolphins.

From the outset, Lilly's accounts of his quest for the "maximum of information from the subjective sphere" had included, alongside reports of childhood memories, sightings of "apparitions," "alien beings," and so forth. If those encounters seemed initially to have been simply by-products of his isolation experiments, they eventually served as a springboard to his exploration of the visionary terrain that Huxley had begun to map. Again Lilly's autobiographical account must be treated with some caution, for it imposes a retrospective clarity on what in all likelihood was a more muddled process. Scattered remarks of his from the 1950s convey some sense of the conflict he evidently experienced as he struggled to reconcile his identity as a professional scientist with a security clearance giving him entry to the highest levels of the science-government nexus and his persona as psychonautical explorer. Responding to a question at one mid-1950s conference regarding the parallels between his own findings on isolation and those reported in the literature about yoga and other Eastern practices, he answered circumspectly that his conversations with Huxley had given him the impression that much of that literature contained assumptions he was not yet prepared to make, namely, those relating to the "mystical character" of the experience. Over time he would abandon such reservations. *The Center of the Cyclone: An Autobiography of Inner Space* (1972), which became a major contribution to the counter-cultural canon inaugurated by Huxley's *The Doors of Perception* (1954), contains an avowedly mystical component.

What is striking in all of this is the proximity of brainwashing to its opposite in Lilly's work. This is clear in the detailed accounts of his experiments in isolation, which in the 1960s eventually incorporated LSD and then ketamine. Their objective, as he articulated it, was a form of self-deprogramming, an escape from the collective hallucination that he now referred to as the "human consensus reality," whose elements were shaped by, among other forces, the mass media of radio, television, motion pictures, and magazines. The altered states he experienced in the tank were visionary in the extreme: they included communication with other intelligences, figures, and "apparitions" not only from his past but also from memories that did not appear to be his own, then dolphins, and ultimately otherworldly beings. In *The Center of the Cyclone* he referred to a hole or interface between "our universe and another one containing alien demonic forms who were pouring into my head from their universe."

Bizarre as those accounts were, in at least one regard they relied on familiar elements, adopting terminology from the this-worldly realm of modern media and technology. Like Penfield, who used the analogy of film for the "flashbacks" experienced by some of his epileptic patients while undergoing surgery, Lilly's reference for the altered states of consciousness he experienced in the tank was film:

> Freed from the effects of gravity, light, and sound in the tank, he was able to study the visual images in a more relaxed state. In the tank he saw continuous motion-picture-like sequences, highly colored, three dimensional, and consisting of, at first, inanimate scenes which later became populated with various strange and unusual creatures as well as human beings. He found that he could change

the content of these internal movies by the self-metaprogramming
methods he had learned in the tank ... under LSD.

Following an extended period of immersion in the tank, he returned
to "the old familiar movies of Earthside scenes and his own memories."

The autobiographical account presented in *The Scientist*, as Charlie
Williams has noted, bears many traces of the midcentury paranoid style,
particularly in its preoccupation with the question of who or what was
controlling Lilly's thoughts and actions. Even as Lilly's research invited
speculation about radical new possibilities for both extracting information from and injecting new information into the human organism, it also
posed questions about the nature and origins of that information. Who
or what, exactly, was the source, the "director" of the visionary trans-
missions reported by Lilly? Throughout his writings the answer took
different forms: the Catholic Church, the Communist police state, the
"human consensus reality," Huxley's "Mind-at-Large." The latter part of
his narrative is marked by increasing uncertainty as to whether the origins of the imagery he experienced in the tank were internal or external.
In those scenes Lilly gives living form to the philosophical thought experiment of the brain in a vat—one in which the answer to the question
of who was controlling the inputs becomes deeply unclear.* He is, in the
end, similar to the scientists in Stanisław Lem's *Solaris* (1961), who end
up being experimented on by the planet they are seeking to investigate.

Conclusion

To recap this remarkable itinerary: in the medical laboratory in which
he embarked upon his scientific training, Lilly tells us, he encountered
for the first time "the fragile, pink, pulsating surfaces of human brains at
operations." During his subsequent years of research, experiences dating
back to his youth that had awakened intuitions of alien intelligence were

---

* "Suspended in warm water, in perfect darkness, Lilly became, you might say, a brain in
a vat." D. Graham Burnett, "A Mind in the Water," *Orion*, May–June 2010, 14.

repressed. He wrote that "he lost contact with the Beings directing his mission. He called such thoughts science fiction. The human consensus reality took over his mind, his body, his brain, his social relations." His turn to perceptual isolation, however, led gradually to the restoration of contact. During the thousands of hours that he spent floating in the tank, the "human consensus reality" took on the contours of a collective hallucination. Meanwhile, the "science fiction" universe of the Beings re-asserted its presence, revealing itself to him as a higher or meta-reality that could be apprehended in the form of "motion-picture-like sequences."

John Lilly's life and career present us with a highly contradictory pic-ture. How can we disentangle the significant role he played in midcentury brain science from the countercultural persona he later adopted, with its strange mixture of heroic self-experimentation and mysticism? Or of either from the Cold War specter of mind control? There are no easy an-swers to these questions. Perhaps, however, there is no real contradiction. From one vantage point it may be seen that Lilly simply traversed the entire spectrum of possibilities opened up by the midcentury sciences of brain and behavior: from the dream of a perfect readout of brain activity, correlating point by point with mental activity; to the political specter of brainwashing; to the sixties phantasmagoria of "Mind-at-Large."* The scientific ferment of that era incubated visionary and to varying degrees fantastical possibilities of many kinds, all of which coexisted closely in Lilly's life and work. Navigating those possibilities, however, came at a cost to the coherence of his work: the explosion so to speak, of all existing disciplinary frameworks.

IF THERE IS A UNIFYING thread running through Lilly's isolation tank experiments, it may be glimpsed in the therapeutic possibilities he ex-plored in connection with them. These had a personal dimension that

---

* That dream is well captured in the following statement of what he aspired to capture: "Total instantaneous electrical activity of the brain, total instantaneous mental life, total instantaneous bodily behavior"—as well as knowledge of the laws governing the relations among these three regions. ("Proposal," May 15, 1952, Box 31, Lilly Papers.)

can be traced to his experience of illness—specifically, to migraine. As it was for Lewis Carroll, migraine was a constant for Lilly. In *The Center of the Cyclone* he described his experience of the syndrome as marked by excruciating pain on the right side of the head for eight hours—a condition he experienced once every eighteen days for more than forty years. Many of the visionary experiences he reported take on a new aspect if we see them in relation to migraine, particularly when we bear in mind the kinship among hallucinations experienced under the influence of strobe, drugs, sensory deprivation, epilepsy, and migraine. Each served as a portal to altered states of being. It seems likely that Lilly's exploration of such portals originated in his experience of migraine.

Support for this suggestion can be found in his late book *Programming and Metaprogramming in the Human Biocomputer*. Published in 1968, the book represents the culmination of a lifetime's worth of research. It is based on a report he submitted to the NIH in the mid-1960s that summarized the results of his research with LSD in the isolation tank (discontinued once the drug was made illegal). It includes a chapter dealing with experiments in "metaprogramming"—essentially, a form of self-guided analysis, or reprogramming—in relation to what he calls the "fixed neurological program" of migraine.* That chapter in turn draws heavily on a section of a research proposal he first drafted in the early 1950s. Here the possibilities of reprogramming are brought to bear on migraine. In the experimental conditions created by LSD and tank, the peculiar disturbances of vision associated with migraine (distortions of the visual field, asymmetries between the right and left eyes, photophobia, scotomas, and "visions") could, Lilly noted, be analyzed and then related to what he called "the long history of the self." In language quite close to that used in his comments on the work of Penfield and Kubie in 1953, he wrote, "'Shocking' (literally) experiences can be 'recalled' and 'relived.'" He cited an experience involving a blow to the head followed

---

* The appendix to *Programming and Metaprogramming in the Human Biocomputer* includes an excerpt from Fyodor Dostoyevsky's *The Idiot*, in which Prince Myshkin recounts the moment of euphoria at the onset of an epileptic seizure, when "his brain seemed to catch fire, and . . . his vital forces were strained to the utmost" (122–25).

by loss of consciousness (a childhood trauma recounted in *The Scientist*). Analysis of that experience revealed the presence of a "long term, apparently built-in, more or less unconscious program." Such passages suggest that Lilly conceived of the neurological syndrome underlying migraine at least partly in terms of a psychoanalytic model in which the malady was related to, or intensified by, repressed memories. As Lawrence Kubie had suggested, the "experimental amniotic fluid" in the tank was a means of accessing the unconscious, in this sense comparable to Penfield's work with his "memory patients."

Migraine, wrote Lilly in the early 1950s, served as a tracer of or spur to self-analysis. As Hughlings Jackson wrote of epilepsy, it was a "natural experiment" whose effects could be reproduced under controlled conditions, then enhanced and studied. But psychoanalysis was only one frame of reference. Though Lilly saw recovery of memory and the past as a crucial element of the resulting effects, he also stressed other possibilities of a decidedly more enigmatic character. If in the early 1950s, when he first encountered the apparitions, he lacked what he called the "transcendence program" necessary to grant them reality, by the end of the 1960s that was seemingly no longer the case. Ultimately the trajectory traced here bears witness to the tangled route by which the vision of the fully programmed brain that was born in the 1950s mutated into its opposite: the fully deprogrammed brain, the brain wiped free of all programs, "the door in the head" now open to new perceptions or programs—to "Mind-at-Large."*

---

* On the fate of Huxley's conception of the brain as reducing valve and of its corollary "Mind-at-Large" in contemporary research in psychiatry, see Nicolas Langlitz, *Neuropsychedelia: The Revival of Hallucinogen Research Since the Decade of the Brain* (Berkeley: University of California Press, 2012). Following a suggestion made by Langlitz, it is possible to see Lilly as oscillating between the dystopian and utopian poles of Huxley's writing that are defined by *Brave New World* and his later novel *Island*.

Conclusion

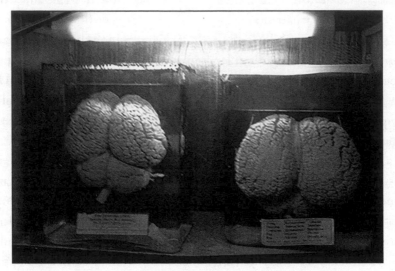

Specimens from the Moscow Brain Research Institute.

REPRINTED WITH THE PERMISSION OF JOY NEUMAYER.

As the curtain came down on the Cold War, journalists announced a startling discovery: tucked away in an obscure research institute in Moscow was a room containing a collection of the brains of members of the former Soviet elite. According to a *New York Times* article in September 1991, the collection, which included the brains of luminaries such as Vladimir Lenin, Josef Stalin, and Maxim Gorky, was stored "in jars and on slides in Room 19 of the Brain Institute in Moscow." The history of the institute, it was later disclosed, dated back nearly to the Soviet Union's beginnings, specifically to the immediate aftermath of Lenin's death following a stroke in 1924. Wishing both to determine

the cause of his death and to secure proof of his status as a world historical thinker, Stalin and the other Bolshevik leaders created an institute
where the necessary studies could be carried out. Eventually the brains
of other Soviet luminaries were added to a collection housed in the institute. Stalin's brain became part of the collection following his death
in 1953 as the result of a massive cerebral hemorrhage. Postmortem investigation revealed that the dictator had suffered from a long-standing
illness that, according to at least one of those present at the time, may
have impaired his decision making. The contents of the Moscow Brain
Research Institute subsequently became exhibits in the bill of indictment
drawn up against the Soviet regime following its collapse. Among the
documents included in "Communist Party on Trial" was the original file
on Lenin's brain, whose status as the basis of a political cult was invoked
by prosecutors as evidence of the elitism of a party that had betrayed its
mission of emancipating the proletariat.

A year before the discovery of this collection was reported, on July 17,
1990, US president George H. W. Bush announced the launch of a major endeavor dedicated to brain research and its public benefits. His announcement affirmed the growing stature of neuroscience at a moment
of major historical transformation. Was it mere chance that the end of
the Cold War coincided with the beginning of the so-called Decade of
the Brain? That a historical epoch stamped by the power of ideology
and its alleged formative influence over consciousness was succeeded by
a period marked by the growing prestige of biology—in particular, the
biology of the brain and therefore, according to many neuroscientists, of
consciousness?

Like the decade that had marked the beginning of the Cold War,
the decade following its conclusion ushered in a new political and cultural
order, heralded by many as a moment marking both the end of history
and the beginning of an era of free markets and democracy. As Bush's
announcement seemed to affirm, the coming of age of neuroscience was
a central feature of the new order. The end of a conflict defined as a war
between starkly opposed ideologies was marked by a search for new uni-
versals. Biology, including the biology of the brain, became one of them.
In the West the heady sense of triumphalism associated with the Soviet

Union's collapse was accompanied by optimistic visions of humanity's future, one element of which was an image of the brain as a resilient, ever more explorable organ. A major campaign dedicated to encouraging public awareness of and dialogue about the medical, ethical, and humanistic implications of neuroscientific research was propelled by the belief that advances in this area promised once and for all to find answers to many long-standing enigmas: mental disease, social ills, even, according to one observer, "the deeper meanings of who we are." The putative universalism of brain science provided one of the foundations of the new postideological, neoliberal world order. The discovery of the Soviet Pantheon of Brains served as an ironic testimonial to this, offering up to Western eyes a tableau of brains afflicted with the disease of a now-defunct ideology.

With the end of the Cold War, physics ceded to the life sciences the mantle of queen of science. The iconic image of twentieth-century science, the atom, was replaced by the image of the brain scan. Aided by the enormous suggestive power of brain-imaging technologies such as functional magnetic resonance imaging (fMRI), claims for the biological basis of consciousness became common. Mind, declared Eric Kandel, is a set of operations carried out by the brain. Assertions such as Kandel's announced the recasting of human life in biomedical terms.

Biology was similarly recast. It was no longer understood simply as fate but as the target of medical or chemical intervention. The twentieth-century view of the adult brain as an essentially hardwired organ gave way to a new appreciation of its ever-evolving, or plastic, nature. Therapeutic breakthroughs were one part of the story. Another was the new possibilities of enhancement and optimization that spilled forth from laboratories and clinics. As the horizons of human self-enhancement expanded, the brain became "invested with ever more extravagant hopes."

Those hopes were fueled by a spate of books charting the remarkable inner landscapes investigated by neuroscientific researchers. Many explored their implications for matters including emotion, morality, creativity, and faith that had long been considered to be the province of fields such as philosophy, religion, humanities, and the arts. Neuroscientists became celebrities. In many fields in the humanities, for instance, the reverence with which intellectual superstars such as Michel Foucault and Jacques Derrida had once been treated was transferred to scientists like Antonio Damasio, Joseph LeDoux, Eric Kandel, and Oliver Sacks. Works of popular neuroscience multiplied prodigiously, feeding a hunger that would have been the envy of midcentury popularizers such as Grey Walter.

What accounts for this hunger? Why has the brain become such an object of fascination? Is it that neuroscience offers the next "grand narrative of human behavior," a new universal at a time of rapidly accelerated change? If so, it is worth paying close attention to this phenomenon, especially since, as is often true of grand narratives, there turn out to be problems with the story. For one thing, some of the most widely publicized claims of neuroscience have been challenged. In the field of mental health care the limited therapeutic payoff of brain-imaging techniques and new medications has led some observers to note that biopsychiatry has failed to deliver on its promises. As was true in the 1950s, the tremendous achievements of contemporary neuroscience have been accompanied by much hype, leading some commentators to identify a new disorder characteristic of the moment: "brain over-claim syndrome."

Meanwhile, a sense of disquiet accompanied the dawning of the era of the brain, particularly as the boundary between treatment and enhancement became ever more blurred. One symptom of this has been the unease provoked by new possibilities for intervention, or the rise of what some refer to as "cosmetic neurology." This was made manifest in a series of reports issued by the President's Council on Bioethics (PCBE), established under President George W. Bush. Featuring titles such as *Beyond Therapy: Biotechnology and the Pursuit of Happiness* (2003), the reports conjured up the specter of tampering with memory or of the psychopharmaceutically administered society first envisioned in Huxley's *Brave New World* in 1932 and later in his warning, in 1946, that humankind stood at the brink of a "Manhattan Project" of human happiness. Now that the Soviet revolution had failed, what was left but that "revolution of body and mind" predicted by Huxley?

The sense of both hope and unease that surrounds modern biotechnology manifests itself in many ways. Take, for instance, EEG, the technique that did so much to open the brain to study in the middle of the twentieth century. Though it has been partly superseded by functional magnetic resonance imaging (fMRI) and other brain-imaging techniques, new uses continue to be found and imagined for it. One of them relates to a mental function that has come under much scrutiny: attention. A seeming epidemic of attention-related disorders has sparked a quest for new forms of attention enhancement. They include biofeedback, which first emerged in the 1960s as a means of training people to control their bodily and mental functions by observing their own brain-wave patterns. It has recently been adopted as a means of treating attentional disorders and improving concentration. It now holds out the possibility of optimizing performance in arenas ranging from the classroom to the workplace. Though it may be true, as the authors of the PCBE report assure us, that in "free societies . . . biotechnical choices are not made by central planners," it also seems true, as Jia Tolentino has suggested, that the self has become one of capitalism's last natural resources and that neuroscientific techniques will play a central role in tapping that resource.

There also remain questions about the role of such techniques within twenty-first-century politics and war. Fiction provides one place to imagine their darker possibilities. Consider, for instance, Adam Johnson's *The Orphan Master's Son* (2012), a novel that depicts modern North Korea as a society in the grip of a regime heavily reliant on Cold War–era techniques such as interrogation and lobotomy to "reform" its citizens. Toward the end of the story a new, seemingly more humane technique is introduced. This is the so-called autopilot, a device that employs feedback and an EEG-like apparatus in order to regulate attitude and behavior. It "works in concert with the mind, measuring brain output, responding to alpha waves." The autopilot's algorithm reads each individual's electronic signature and then applies the data in the service of a vision of "the self and state, coming closer to one another."

That the specter of interrogation hangs over Johnson's novel is no accident. As it turned out, the new age of enhancement also encompassed such practices as well, in the form of "enhanced interrogation." Ten years after the Soviet Union's collapse, the shock of 9/11 triggered an enormous expansion of the national security apparatus paralleling that of the early

Cold War. Within a few years, stories began circulating in the press about unsavory aspects of the war on terror that eerily evoked that earlier conflict. A veil was gradually pulled back on an all-but-forgotten history of US involvement in torture dating back to the 1950s.

The threads of that history eventually led back to the CIA's 1963 classified interrogation manual, *KUBARK*, and behind it to the "Manhattan Project of the mind" launched during the Cold War's first decade. The practices then devised resurfaced in reports about the practices used across a global network of detention centers. Guantánamo became symbolic of a return of the repressed, a return that forced American society to confront the amnesia that surrounds much of its own recent history, not least that of the relationships among secrecy, democracy, and torture. Reports on the experiences of detainees there document the combination of techniques of over- and understimulation—including, among others, strobe and sensory deprivation—that became part of the arsenal of interrogation.* Such revelations serve also as reminders of the now-seventy-year-old history of the alliance between the behavioral and psychological sciences and the national security state. As if to confirm Huxley's speculations about a "Manhattan Project" of human happiness, a central player in the story was University of Pennsylvania professor of psychology Martin Seligman, best known for his writings on "authentic happiness" but also for his ideas about learned helplessness, ideas that featured prominently in enhanced interrogation. The APA report issued in 2016 in response to the torture scandal laid out a meticulous chronicle of the discipline's relation to the military establishment dating back to the beginning of the Cold War as part of a long-delayed reckoning with the ethical transgressions of some members.

These developments underscore the fact that, as in the 1950s, the brain has become central to many of the major contestations of our time. Turning from world events back to neuroscience, this is manifest in the

---

* Strobe has also been weaponized by hackers who have used Twitter to send videos of flashing lights in order to trigger seizures in members of the Epilepsy Foundation. See Manny Fernandez, "Epilepsy Foundation Was Targeted in Mass Strobe Cyberattack," *New York Times*, December 16, 2019.

field's ambition to establish itself as a new universal serviceable for a post-ideological age. This ambition has been accompanied by repeated efforts to slay what it regards as old dragons. The list is long and includes religion, Cartesian dualism, Freud, antipsychiatry, and others.

Such efforts have not gone unchallenged. As the essayist Curtis White has observed in connection with neuroscience, all claims to move beyond ideology are pure ideology. As neuroscience extends its claims over consciousness, many observers note with dismay—and no small amount of defensiveness—the resulting encroachment into terrain long seen as the preserve of theologians, philosophers, humanists, and artists. The decade of the brain was accompanied by numerous expressions of this. Writing about contemporary brain science's relation to the master narratives of the twentieth century, Tom Wolfe declared in 1996, "Freudianism and Marxism—and with them, the entire belief in social conditioning—were demolished so swiftly, so suddenly, that neuroscience has surged in, as if into an intellectual vacuum."

However overstated, such views captured the magnitude of changes ushered in by the rise of the contemporary neurodisciplines. But all but absent in this "surge" was any kind of deeper historical perspective. End-of-history pronouncements accompanying the conclusion of the Cold War were also adopted by the neurosciences, which sought to banish their precursors to the prescientific past. Over time a certain narrative of brain science was naturalized; stripped of history, it became quasi-mythic, endlessly recycling certain well-traveled tales that serve as markers in a story of steadily advancing scientific progress out of the darkness of the past. Yet in its assertive present-mindedness, contemporary neuroscience finds itself continually disturbed by uncomfortable reminders of that past. Here, too, there is a kind of return of the repressed, to again use the language of one of the most stubborn of those precursors, Sigmund Freud.

Take, for instance, the writings of the neuroscientist Antonio Damasio, who has sought to establish a central place for emotion in his account of the thinking mind. In doing so, he is trying to undo what he calls "Descartes' error": to overturn the half-millennium-old division between immaterial brain and physical body and replace it with a corporeal, emotional brain. The Cartesian idea of the disembodied mind has, he wrote, had many long-enduring effects, among them that recurring phantasm of philosophical thought experiments: the brain in a vat. Lacking inputs from the body, including emotional inputs, "such a brain would not have a normal mind." The consequences of Cartesian dualism, according to Damasio, have been grave: "Mind has been left out of medicine until quite recently, left to religion and philosophy." This has resulted in an impoverished conception of the suffering self, a kind of empty space that has been partly filled by alternative medicine (Scientology can stand in as one example of this). In place of the dualist Descartes, Damasio has embraced the monist Baruch Spinoza, the subject of his 2003 book *Looking for Spinoza*. Most figures in the field seemingly now regard themselves as mind-brain monists. But philosophers such as Ian Hacking and others cast some doubt on the monist claims of contemporary neuroscience. Siri Hustvedt, meanwhile, has observed that the brain has become such a fetish object that its status as an embodied organ is continually forgotten. Mental processes such as reasoning are, as she put it, often still treated as if they took place in a bodiless organ, "a bunch of neurons in a vat going about its business alone."

Or consider the present-day relationship between neuroscience and psychoanalysis. Freud provided one of the most powerful attempts to fill the empty space left by mind-brain dualism.* In the second half of the twentieth century, however, following the psychopharmaceutical revolution of the 1950s, a new biochemical model of mental illness gained ascendance. The result was what some describe as a gradual flattening out of the Freudian depth model of the psyche that dominated the first half of the twentieth century. Since the 1980 edition of *DSM*, psychiatry has increasingly identified itself as "biopsychiatry," a brain science. In accounts of its rise, those who do not fit comfortably within the emerging

---

* The acolytes of modern brain science always found it easier to make room for Pavlov than for Freud in their models. In a debate with Lawrence Kubie in the 1940s, the cybernetician Walter Pitts compared the unconscious to the appendix: useless but prone to disease. See Siri Hustvedt, *A Woman Looking at Men Looking at Women: Essays on Art, Sex, and the Mind* (New York: Simon & Schuster, 2016), 256.

narrative are often written out of the story, treated as wrong turns or dead ends on an arc that culminates in the era of Prozac. This is true equally of Freud and of antipsychiatrists such as Laing and Foucault.

To some critics the rise of biopsychiatry has enthroned a model that comes close to being "mindless," one that repeats many of the errors of the age of somatic psychiatry that spawned antipsychiatry in the first place. This tendency has in some respects been compounded by the excesses of those very critics. The antipsychiatrists' attacks on what they saw as the reductive and unethical stance of brain science eventually hardened into a new orthodoxy in which the "mindlessness" of midcentury brain science was replaced by a model of mind that was "brainless," a model that made it all too easy for its opponents to write them off as scientific naifs. What began as an intellectually fruitful exchange between two tendencies that were perhaps not as incompatible as they at first seemed degenerated into sterile and unproductive antagonism and mutual incomprehension.

Yet even as biopsychiatry celebrated its success in vanquishing Freud, leading neuroscientists such as Antonio Damasio, Eric Kandel, and Joseph LeDoux tried to find common ground with psychoanalysis and its still powerful model of the mind. According to Kandel, the richness of the ideas of his childhood idol Freud was marred by the fact that they paid little attention to biology, and that neglect was repaid in kind by scientists rapidly shifting to a biological model of consciousness. When attention was paid to the unconscious factors of mental life, it was the visceral or physiological rather than the Freudian unconscious, an entity seen as hopelessly unscientific. A small handful of midcentury figures harbored hopes of a future synthesis of the neuro- and psysciences. Although historians judge those hopes to have remained largely unfulfilled, efforts to forge such a synthesis persist. One expression of this is the renewal of interest in the potential for using psychedelics to treat mental illnesses such as depression and PTSD. Building on a largely forgotten body of work dating from the 1950s, this approach uses psychedelics to "depattern" the brain's normal conditioned responses, enabling the patient to confront debilitating fears. The deactivation of inhibiting mechanisms puts different brain centers into communication in a way that may give rise to hallucination and memory. As it did for scientists in the 1950s, the study of hallucination seems to help crack open some of the secrets of the relationship between mind and brain.

Further evidence of neuroscience's selective relationship to its own past is evident in the treatment of some of the key episodes of that past. As we saw in earlier pages, the 1950s included several such episodes that helped establish its claim to be one of the most revolutionary decades in the field's history. Among them was the surgical procedure performed by William Scoville on H.M. that led to the destruction of his memory. Though the ethical issues raised by the case of H.M. are now no longer ignored, they continue to be treated somewhat gingerly. Or take the controversy that surrounds the research on sensory deprivation (SD) by Donald Hebb, a major architect of the neuroscientific paradigm. In the wake of the torture scandal that erupted in 2004, some authors implicated Hebb and his investigations into SD in the practices codified in *KUBARK*. Hebb has since been exonerated, though a form of guilt by association seems to cling to him relating to the work of his Montreal colleague the psychiatrist Ewen Cameron, whose experiments in "psychic driving," which included SD, became emblematic of the midcentury assault on memory in the name of cure. The reckless if accidental destruction of H.M.'s memory at Scoville's hands was in Cameron's case transformed into a methodical, systematic form of man-made amnesia designed to create a blank slate on which new, "healthy" patterns of thinking could be imprinted.

A notable feature of these uncomfortable reckonings is that the place of sensory deprivation within the history of neuroscience has been largely forgotten. But research into SD engaged many leading scientific minds of the 1950s. Midcentury practitioners explored its potential as a means of treating disorders such as migraine. They also explored the implications of research into its effects on healthy subjects. Such research suggested that conventional notions of the stability and integrity of the self were quite erroneous—an issue of no small significance at a moment in which the speeding up of history seemed to have brought about a "fundamental disturbance of man's relation to his environment." Last but not least, for Hebb himself sensory deprivation played a crucial role in helping break the grip of behaviorism and its ban on the scientific study of consciousness over his field. In demonstrating that the mind deprived of external stimulation generated internal imagery that he described as "motion-picture like hallucinations," it contributed one of the central findings of the 1950s: the new picture of the brain as continuously active organ. SD was central to the transformation of scientific understanding of the central nervous system that took place during the 1950s.

How, we may wonder, can such episodes be properly integrated into the story of modern brain science? Only with difficulty, it seems. Given the field's selective relation to its own past, it is ironic that one of the foundational elements of neuroscience's ascendance has been the study of memory. For it is in relation to this crucial faculty that some of the greatest disquiet about contemporary developments has been expressed. Since the turn of the century, neuroscientists have probed new possibilities for both enhancing and modifying memory. Some of these possibilities are deeply unsettling, particularly those relating, for instance, to the conduct of war. Just as the Cold War served as a laboratory for testing new approaches, so, too, has the war on terror. The epidemic of brain trauma and of post-traumatic stress disorder (PTSD) experienced in connection with that never-ending conflict has prompted figures such as the neuroscientist Joseph LeDoux to pose the question: Should memory be off limits as a therapeutic target? Might it be therapeutically (and perhaps also strategically) advisable to dose soldiers with beta-blockers prior to their entering combat in order to prevent the encoding of disturbing memories? It is precisely such possibilities that the President's Council on Bioethics highlighted in posing questions about the implications of treating memory as a malleable substrate. Contemplating such possibilities, its authors, who are perhaps guilty of overstating the case, invoked the warnings of Huxley, Orwell, and Arendt: "Nor can we forget the central place of manipulating memory in totalitarian societies, both real and imagined, and the way such manipulation made living truthfully—and living happily—impossible." As such statements suggest, the specter of brainwashing, understood above all as the erasure of memory, continues to haunt present day brain science and liberal democracy.

◆ ◆ ◆

THE BRAIN'S EMERGENCE AS A talismanic object of modern science has invested this most charismatic of the body's organs with hope, desire, and dread. Many of these emotions are bound up with images of brain scans, which seem to contain urgent messages about mankind's present and future. Translation of those messages has become the task of an ever-growing crowd of specialists. But if the story of the modern brain attests to the suggestive power of the images produced by modern scanning technology, images that cast an undeniable spell, it also attests to the continuing opacity of these images, which remain despite everything rather inscrutable. This raises the question: What role in the task of elucidating the brain's secrets will be played by fields that neuroscience has sought to displace—philosophy, religion, the arts?

As Damasio has acknowledged, the mind-body split that still remains only partially overcome in the modern neurodisciplines has left an opening for the approaches of competing disciplines. These include that most ancient of approaches to what it means to be human, namely storytelling. The modern history of the brain provides many reminders of the possibilities of storytelling that surround this organ, especially stories relating to patients and to the fantastical inner landscapes their illness narratives open up. No doubt this is part of the reason that the midcentury scientists central to this book adopted Lewis Carroll as their patron saint, for the stories of the migraine-afflicted Oxford don seemed to capture something vital about those inner scenes in a way that few scientific texts could.

One example of this is a key text of the modern history of the brain, the Russian neuropsychologist A. R. Luria's *The Mind of a Mnemonist* (1967). It begins with an epigraph from *Alice's Adventures in Wonderland* and continues, "Together with little Alice we will slip past the smooth, cold surface of the looking glass and find ourselves in a wonderland, where everything is at once so familiar and recognizable, yet so strange and uncommon." The book that follows tells the case history of his patient S., a man endowed with astonishing powers of recall. Through a relationship stretching over decades, Luria established that S.'s capacity for memory was based on an unusually figurative, synesthetic form of mental process. His remarkable capacity to store and reproduce numbers, names, and other data was linked to an ability to "picture" such information that extended to other functions as well. As S. put it, "If I want something to happen, I simply picture it in my mind." Like an Indian yogi or a biofeedback subject, he could regulate his own autonomic processes. For instance, he achieved perfect control of his heart rate and body temperature simply by "picturing" himself running after a train, lying in bed, or holding his hand to a flame. In one experiment in which he imagined that a stroboscopic lamp was flashing in his eyes, his EEG showed a "distinct depression of the alpha waves." For Luria the inner world opened up by S. was so amazing that it continually evoked the experiences of Alice.

Luria's ability to enter his patient's inner world made him one of the founders of a medical-literary genre that has been adopted by many as a model for their "clinical tales." Take, for instance, Oliver Sacks, another of the major chroniclers of the story of the modern brain. In his hands this is marked by a commitment to integrating the experience of the patient into medicine. The generosity that marks his tales of recovery from or adaptation to serious illness is evident as well in the way he continually repays his intellectual and scientific debts to his predecessors. One such figure is Wilder Penfield, a constant presence in Sacks's writings on epilepsy, migraine, hallucination, and memory. This remained true even when he acknowledged the limitations of Penfield's belief that memories are stored like frames of a film: "We now know that memories . . . are transformed, disassembled, reassembled, and re-categorized with every act of recollection" (though some traumatic memories, he adds, remain vivid and relatively fixed through life, just as Penfield seemed to show).

But although Sacks pays homage to figures such as Penfield and the midcentury cohort that helped usher in a new era of brain science, he also notes their blind spots, their inability to do justice to the complexity of mental life or to the nervous system's capacity for adaptation to injury. In his memoir he noted the recent emergence of a new vision of the brain that takes account of the damaged brain's resilience, its ability to reassign functions to different areas. Beyond that, however, what remained most significant for him in his interactions with patients was not just function but the meaning they attached to their experiences of illness—the ways in which their illnesses became woven into complex personal narratives. What he called the Proustian nature of reminiscence eluded the "cybernetic" model central to Penfield's work.

Underlying this was a related concern about the ethical implications of some of Penfield's case histories, which showed that "the removal of the minute, convulsing point of cortex, the irritant focus causing reminiscence, can remove *in toto* the iterating scene, and replace an absolutely specific reminiscence or 'hyper-mnesia' by an equally specific oblivion or amnesia." Paradoxically, the stability of memory that Penfield believed he had demonstrated seemed to open up something like the very opposite. To Sacks's mind it raised the frightening possibility of a "neurosurgery of identity (infinitely finer and more specific than our gross amputations and lobotomies)." What might be lost in neurosurgical operation? he asked. Memory; but perhaps also the meaningful, even pleasurable or addictive, qualities of epilepsy that were described by Dostoyevsky, which, he pointedly noted, Penfield never mentioned in his writings.

Like Freud and Luria, Sacks stressed the value of case histories as a source of new knowledge, with patient teaching doctor as much as doctor treating patient. That view was shared by Penfield, who encouraged his memory patients to speak on the operating table. Yet Penfield ultimately saw only randomness in his patients' "reminiscences." Sacks, on the other hand, stressed the need for "analysis in depth," a process in which free association might uncover deeper meaning. Whereas Penfield viewed the flashbacks produced by stimulating a point in the cortex as lacking meaning, the case of Sacks's patient Mrs. O'C., who experienced vivid reminiscences in connection with the onset of epilepsy, offered what he regarded as evidence of the unconscious, repressed, highly charged nature of the memories brought forth under such conditions. Some of Mrs. O'C.'s reminiscences, he added, had a close kinship with Dostoyevsky's experience of "happiness" associated with a seizure. In his patient-centered narratives Sacks offers one way to bridge the divide that marks the history of the neuro- and the psysciences.

Yet it must be asked whether Sacks's essentially optimistic vision is in the end not too optimistic. Just how resilient, after all, is the brain? Conversely, does the brain's much-touted plasticity not open up troubling prospects? War again seems to be the great laboratory: the possibility now contemplated of not simply treating soldiers' psychic wounds through drugs intended to weaken traumatic memories but of optimizing or enhancing their battlefield performance by means of fanciful devices such as the NeuroSky MindWave, a headset for monitoring brain waves. Though Sacks's historical vision ranges widely across the medical and scientific past, much of his chronicle of that past treats it as though it existed in a social and political vacuum. This leaves a blind spot in his own writings: the darker possibilities that hover around his largely optimistic story remain in the background. As Curtis White has put it in his critical analysis of contemporary neuroscience, the brains depicted by Damasio and company seem like such cheerful, bourgeois brains.

As the deeply worried tone of the PCBE's discussion of memory makes clear, the paranoid style that haunted the origins of modern brain science in the 1950s has made a comeback in the early twenty-first century.* A major examination of neuroscience published in 2013 was titled *Brainwashed*, and its survey of the contemporary uses and abuses of neuroscience in many fields included a discussion of neuromarketing that harkened back to the anxious reflections of Vance Packard. Other critics raise similar concerns in posing the question: How does today's neuro-hype mesh with the ideology of capitalism? As in the 1950s, a crowd of specialists has once again gathered, intent on getting inside our heads, and this conjures up unnerving possibilities. The sense that the contents of our heads are no longer private, or even our own, is well captured by the question raised by the digital activist Eva Galperin: "It used to be that if you wanted to look into somebody's brain, you needed to talk to them . . .

---

* The specter of the Manchurian Candidate has also been a meme of every presidential campaign since 2000, most recently in the charge that Trump was a "Siberian Candidate." In this as in other respects, the United States continues to grapple with the protean legacy of the Korean War. Paul Krugman, "Donald Trump, The Siberian Candidate," *New York Times*, July 22, 2016.

[or] torture them." But nowadays, "Who needs interrogation when a cell phone is essentially the inside of your brain?"

There are, it seems, many reasons why brainwashing remains "the world's favorite conspiracy theory." In its melodramatic way, this construct expresses the anxiety and disorientation, the "agency panic," felt by many modern individuals amid the crises of contemporary society. Nowadays brainwashing has become a kind of vernacular term for social media effects and the massive erosion of privacy associated with those effects. In our state of distracted half-knowledge, the nagging sense that each of us is afflicted with attentional or memory-related disorders, the reliance on such constructs may help us feel that we are "in the know." As such it has become part of the folklore of modern brain science.

If the brain has in many ways emerged as the scientific object par excellence, it is, as this book has shown, a thoroughly cultural object, deeply enmeshed in wider historical developments, hopes, and anxieties. The emblematic figure of the modern history of brain science is Henry Molaison or, as he was known for decades, H.M. This is first of all because his story serves in many respects as that history's origin story, which took shape around the scientific study of memory that was made possible by his drastic impairment. But it is also because he seems to embody that condition of man-made, and at least partly organized, amnesia that has been part of US history, and part of the history of neuroscience, since the time his memory was destroyed. It is the aim of this book to undo some of that amnesia.

Acknowledgments

I have gotten more pleasure from the process of writing this book than from any book I've previously written. In many ways I wish the writing of this book could continue indefinitely, such has been the pleasure it has brought me. Alas, that is a luxury I don't have!

The process began serendipitously with a chance discovery of Grey Walter's *The Living Brain* at my local secondhand bookstore (Unnameable Books—thanks!). As it did for many of its midcentury readers, this book opened up for me a fascinating panorama: of discoveries in the field of brain science and of the memorable figures, including Walter himself, who made those discoveries. Having fallen under the spell of this book, I decided to learn more about its author and the scientific revolution to which he contributed, as well as about the historical circumstances in which that revolution took place.

This process introduced me to, and put me in the debt of, many other scholars who have explored aspects of the story presented here. In several cases they have become good friends, and that, too, has been part of the great pleasure of writing this book. One strand of the story can be traced back to a stay at the Max Planck Institute in Berlin and to my introduction to the formative work of Cornelius Borck on the history of EEG. Another goes back to a conference on the topic of "brainwashing" that I co-organized with Stefan Andriopoulos at Columbia University a decade ago. It was through that highly fruitful collaboration that I also met and befriended a handful of other scholars: the late Alison Winter, Tim Melley, Rebecca Lemov, and Stefanos Geroulanos. Their work has been profoundly inspirational for me, and I would like to thank all of them, especially Stefan.

At a later stage in the process, I was very fortunate to come into contact with a group of younger scholars whose work has been quite important for the final direction my book has taken—as I'm sure it will be for future studies in this area. Danielle Carr, Yvan Prkachin, Jeffrey Matthias, Marcia Holmes—I would like to thank each of them in turn for their great generosity, advice, and excellent humor.

For help in ways both big and small, thanks are also due to Joan Talley Dulles, who spoke with me about her brother Allen Macy Dulles and graciously granted permission to cite from her mother's journal entries about him; Robert Lifton, with whom I spent an afternoon discussing his experiences interviewing returning POWs in 1953; Richard LeBlanc of the Montreal Neurological Institute, who gave me a guided tour of the operating theater where Wilder Penfield performed neurosurgery; Gyorgi Buzsaki; Siri Hustvedt; Joe Ledoux; Jesse Prinz; Alec Nevala-Lee; Marcia Holmes and Charley Williams; and Andras Kisery.

For opportunities to present my work, I owe thanks to Carolyn Dean, George Makari and the Richardson History of Psychiatry Seminar, Hannah Zeavin and Jeffrey Matthias, Daniel Pick, James Kennaway, and Stefanos Geroulanos. Much of the research for the book was aided by grants from the CUNY Research Foundation and support from the Rifkind Seminar, where I led a seminar on the "History of Emotions" that was also formative for the book.

Special thanks as well to those people who read parts of the manuscript or drafts of chapters at various stages of their development: Elliot Jurist, Sander Gilman, Ferenc Mechler, Matteo Carandini, Kabir Dandona, Franny Nudelman, Marcia Holmes, Alexandra Bacopoulos-Viau, Rachel Kravetz, Joe Demarie, Willie Neumann, Mikhal Dekel, Patrick Kilian, Andras Kisery, Vaclav Paris, Robert Higney, Harold Veeser, and Gary Wolf, who first introduced me to the "brain-in-a-vat" thought experiment and offered valuable feedback at several junctures along the way.

Numerous archivists and librarians provided crucial help: especially Tim Noakes at Stanford Special Collections, but also staff at the Library of Congress, the Rockefeller Archive Center, the New York Public Library Manuscripts Division, the National Archives, the Hoover Center, MIT,

McGill University, the Wisconsin Historical Society Archives, the Eisenhower Library, and the Weill Cornell Medical Library Archives.

Last and certainly not least, I would like to extend my deepest gratitude and warmest thanks to my students, both those who served as sounding board for my work and those who provided material assistance, whether as researchers or as readers and annotators: Christian Camarero, Ye Zhou, Kailee Neal, Susan Evans, and Brianna Lafoon. I'm especially grateful to Sarah Jenkins Fox, who was a fantastic research assistant and who located some real treasures for me.

My brother, Stefan, lent critical help with the images in the final stages, and I thank him from the bottom of my heart.

Final special thanks to my publisher, Jonathan Jao, and to my agent, Richard Abate, and of course to my family, especially to Nicolas for his help with the images.

Notes

Preface: Experiments on Consciousness

x   *On the Origin of Species*: Gordon M. Shepherd, *Creating Modern Neuroscience: The Revolutionary 1950s* (New York: Oxford University Press, 2010), 164.

xi   "conscious experience": Herbert H. Jasper, "Introduction," in *Brain Mechanisms and Consciousness: A Symposium Organized by the Council for International Organizations of Medical Sciences*, edited by Edgar E. Adrian, Frederic Bremer, and Herbert H. Jasper (Oxford, UK: Blackwell Scientific Publications, 1954), xv.

xi   "We live": "Exploring the Brain of Man: In Search of the Prevention and Cure of Neurological Disorders," National Institutes of Health (1960).

xii   "beginning, so to speak": Alfred E. Fessard, "Mechanisms of Nervous Integration and Conscious Experience," in Adrian et al., *Brain Mechanisms and Consciousness*, 201.

xii   "explorable brain": Rachel Elder, "Speaking Secrets: Epilepsy, Neurosurgery, and Patient Testimony in the Age of the Explorable Brain, 1934–1960," *Bulletin of the History of Medicine* 89, no. 4 (Winter 2015): 761–89; Shepherd, *Creating Modern Neuroscience*.

xiii   "are re-enactments ": Wilder Penfield, "Studies of the Cerebral Cortex of Man: A Review and an Interpretation," in Adrian et al., *Brain Mechanisms and Consciousness*, 293.

xiii   By the 1950s, Penfield: Katja Guenther, *Localization and Its Discontents: A Genealogy of Psychoanalysis and the Neuro Disciplines* (Chicago: University of Chicago Press, 2015), 174.

xiii   "like a strip of film": Penfield, "Studies of the Cerebral Cortex of Man," 295.

xiii   "previously edited": Ibid., 301.

xiv   stroboscopic flicker: Grey Walter, in "Group Discussion," in Adrian et al., *Brain Mechanisms and Consciousness*, 306.

xiv   "the latent emotional impact": Lawrence S. Kubie, in "Group Discussion," in Adrian et al., *Brain Mechanisms and Consciousness*, 309.

xiv   Those experiments, he claimed: Lawrence S. Kubie, "Psychiatric and Psychoanalytic Considerations of the Problems of Consciousness," in Adrian et al., *Brain Mechanisms and Consciousness*, 467.

xv   "differently conscious": Fessard, "Mechanisms of Nervous Integration and Conscious Experience," 201.

xv   Hebb discussed his own: Donald Hebb, "The Problem of Consciousness and Introspection," in Adrian et al., *Brain Mechanisms and Consciousness*, 415–17. See also Caro W. Lippman, "Certain Hallucinations Peculiar to Migraine," *Journal of Nervous and Mental Disease* 116, no. 4 (1952): 346–51; John Todd, "The Syndrome of Alice in Wonderland," *Canadian Medical Association Journal* 73, no. 9 (1955): 701–4.

xvi   "Hypnotism or brainwashing": Robert Morison, "General Discussion," in Adrian et al., *Brain Mechanisms and Consciousness*, 494.

xvi   "It is man's consciousness": "Brainwashing: The Communist Experiment with Mankind," April 19, 1955, NSC Papers, Dwight D. Eisenhower Presidential Library, Abilene, KS.

xvii "The work we have done": Donald Hebb, preface, in *Sensory Deprivation: A Symposium Held at Harvard Medical School*, edited by Philip Solomon et al. (Cambridge, MA: Harvard University Press, 1961), 6.

xvii the American Dream reached: Bruce Cumings, *The Korean War: A History* (New York: Modern Library, 2011), 67.

xvii "experimented with": David Halberstam, *The Fifties* (New York: Fawcett Books, 1993), ix–x.

xviii "relations with the outer world": David Riesman, *The Lonely Crowd: A Study of the Changing American Character* (New Haven, CT: Yale University Pres, 1950), 21–22.

xviii "human personality is not": Harold Wolff, SIHE Annual Report, 1957, Box 6, folder 15, Harold Wolff Papers, Medical Center Archives, New York Presbyterian /Weill Cornell, New York.

xviii "the importance of Hebb's": Robert S. Morison, diary, January 25, 1954, Rockefeller Foundation (hereafter RF), RG 12, 335, Rockefeller Archive Center, New York (hereafter RAC).

xviii "an entity that exists": D. O. Hebb, "The Motivating Effects of Exteroceptive Stimulation," *American Psychologist* 13, no. 3 (1958): 110.

xix   "brainwashing": Hannah Arendt, "Cybernetics," 1964, Speeches and Writings File, 1923–1975, Essays and Lectures, Hannah Arendt Papers, Library of Congress, Washington, DC (hereafter LOC).

xix   best of its kind: Robert S. Morison, diary, August 28, 1953, RF, RG 12, 336, RAC.

xx   those developments laid: Shepherd, *Creating Modern Neuroscience*, 9; Jack D. Pressman, *Last Resort: Psychosurgery and the Limits of Medicine* (Cambridge, UK: Cambridge University Press, 1998), 12; Ian Hacking, introductory essay, in Thomas S. Kuhn, *The Structure of Scientific Revolutions*, 50th anniversary edition (Chicago: University of Chicago Press, 2012 [1962]), ix; Jonathan D. Moreno, *Mind Wars: Brain Research and National Defense* (New York: Dana Press, 2006).

xx   "Manhattan Project of the Mind": Alfred W. McCoy, *A Question of Torture: CIA Interrogation, from the Cold War to the War on Terror* (New York: Holt, 2006), 7; Jane Mayer, *The Dark Side: The Inside Story of How the War on Terror Turned into a War on American Ideals* (New York: Anchor, 2009); Rebecca Lemov, *World as*

Laboratory: Experiments with Mice, Mazes, and Men (New York: Macmillan, 2013); Moreno, *Mind Wars*.

xx   social change at home: Eric Hobsbawm, *The Age of Extremes: A History of the World, 1914–1991* (New York: Vintage, 1994).

xx   hidden forces that sought: Abbott Gleason, *Totalitarianism: The Inner History of the Cold War* (New York: Oxford University Press, 1995), 103.

xx   the entry of what: Richard Hofstadter, *The Paranoid Style in American Politics* (New York: Vintage, 1963), 3.

xxi   Kubie expressed hope: Kubie, "Group Discussion," 467.

xxi   "master discipline": Pressman, *Last Resort*, 36–37.

xxi   accepting its usage: Kubie, "Psychiatric and Psychoanalytic Considerations of the Problem of Consciousness," 422, 444.

xxii   the brain compensated: Oliver Sacks, *Hallucinations* (New York: Vintage, 2012), 34–44.

xxii   "Decade of the Brain": George Bush, "Project on the Decade of the Brain," Presidential Proclamation 6158, July 17, 1990, https://www.loc.gov/loc/brain/proclaim.html.

xxii   "the rising star": Elaine Showalter, quoted in Max Stadler, "The Neuromance of Cerebral History," in *Critical Neuroscience: A Handbook of the Social and Cultural Contexts of Neuroscience*, edited by Suparna Choudhury and Jan Slaby (London: Wiley-Blackwell, 2012), 138.

xxii   "the seat of reason": Christof Koch, "The Footprints of Consciousness," *Scientific American Mind* 28, no. 2 (March 2017): 52–59; Mark Jackson, *The Age of Stress: Science and the Search for Stability* (Oxford, UK: Oxford University Press, 2016).

xxii   "master organ": Shepherd, *Creating Modern Neuroscience*; Stadler, "The Neuromance of Cerebral History"; Yvan Prkachin, "'The Sleeping Beauty of the Brain': Memory, MIT, Montreal, and the Origins of Neuroscience," *ISIS* 112, no. 1 (March 2021): 22–44.

xxiii   the most revolutionary decade: Shepherd, *Creating Modern Neuroscience*, 164.

xxiii   a paradigm shift: Kuhn, *The Structure of Scientific Revolutions*.

xxiii   "groping efforts": H. W. Magoun, "The Ascending Reticular System and Wakefulness," in Adrian et al., *Brain Mechanisms and Consciousness*, 1.

Chapter 1: Brain Science at the Dawn of the 1950s

3   The scene in: W. Grey Walter, *The Living Brain* (London: Gerald Duckworth, 1953), 120.

6   "the brain remained": Ibid., 40–42.

7   under the Nobel laureate Edgar Adrian: Ibid., 53; Gordon M. Shepherd, *Creating Modern Neuroscience: The Revolutionary 1950s* (New York: Oxford University Press, 2010), 45; Rhodri Hayward, "The Tortoise and the Love-Machine: Grey Walter and the Politics of Electroencephalography," *Science in Context* 14, no. 4 (2001): 615–41; Cornelius Borck, *Brainwaves: A Cultural History of Electroencephalography* (New York: Routledge, 2018), offers an invaluable overview of the history of EEG.

7 And it was Berger: Walter, *The Living Brain*, 50–51.

8 "black box" of the brain: Shepherd, *Creating Modern Neuroscience*; Nicolas Langlitz, *Neuropsychedelia: The Revival of Hallucinogen Research Since the Decade of the Brain* (Berkeley: University of California Press, 2012), 104, 194.

8 the foundational concept of feedback: Norbert Wiener, *The Human Use of Human Beings* (New York: Da Capo Press, 1950), 33; Andrew Pickering, *The Cybernetic Brain: Sketches of Another Future* (Chicago: University of Chicago Press, 2010).

8 "the first act of creation": Walter, *The Living Brain*, 26.

9 the radical change taking place: Ibid., 83.

9 sometimes controversial methods: Hayward, "The Tortoise and the Love-Machine," 629.

9 Not least among the contributions: Jack D. Pressman, *Last Resort: Psychosurgery and the Limits of Medicine* (Cambridge, UK: Cambridge University Press, 1998), 356–59.

9 "Precise knowledge": A. R. Luria, *The Man with a Shattered World: The History of a Brain Wound* (Cambridge, MA: Harvard University Press, 1972), 23.

9 "man's tardy recognition": Walter, *The Living Brain*, 255.

9 "no more sacred": Quoted in Elliot S. Valenstein, *Great and Desperate Cures: The Rise and Decline of Psychosurgery and Other Radical Treatments for Mental Illness* (New York: Basic Books, 1986), 225–26.

9 Roland Barthes's short essay: Roland Barthes, "The Brain of Einstein," in *Mythologies*, translated by Annette Laver (New York: Noonday Press, 1957), 68–70; see also Borck, 253–54.

11 extrascientific interests: Hayward, "The Tortoise and the Love-Machine," 616.

11 "the image and symbol": W. Grey Walter, "My Miracle," http://cyberneticzoo.com /wp-content/uploads/2010/01/Grey-Walter-My-Miracle.pdf, 45.

12 "Rosetta Stone": Flo Conway and Jim Siegelman, *Dark Hero of the Information Age: In Search of Norbert Wiener* (New York: Basic Books, 2005), 282.

12 He saw himself: Robert S. Morison, diary, June 22, 1948, RF, 6.1, RAC.

12 "singularly Mephistophelean beard": Robert S. Morison, diary, July 11, 1949, RF 6.1, 2.1-19, 174, RAC.

12 "a nightmare TV or movie set": J. Maier, diary, March 30, 1955, RF 6.1, 2.1-19, 174, RAC.

13 "Grey Walter had": Conway and Siegelman, *Dark Hero of the Information Age*, 282.

13 "no immodest exposure": Walter, *The Living Brain*, 11.

13 "Nor shall we ever know": Ibid., 28.

13 "was a supreme event": Ibid., 35.

14 "The experience of homeostasis": Ibid., 39.

14 "held literally in the midst": Ibid., 86.

15 "provide the mirror": Ibid., 63.

15 the main patterns of brain activity: Ibid., 87.

15 surgical intervention: Ibid., 56.

15 "mosaic of aerial photographs": W. Grey Walter, quoted in John Geiger, *Chapel of*

Extreme Experience: A Short History of Stroboscopic Light and the Dream Machine (Brooklyn, NY: Soft Skull Press, 2003), 17.

16   "moving picture" of brain activity: Walter, *The Living Brain*, 61–63, 258–59; Cornelius Borck, *Brainwaves: A Cultural History of Electroencephalography* (New York: Routledge, 2018), 259–69; Jimena Canales, "A Number of Scenes in a Badly Cut Film: Observation in the Age of Strobe," in *Histories of Scientific Observation*, edited by Lorraine Daston and Elizabeth Lunbeck (Chicago: University of Chicago Press, 2011), 254, n. 95.

16   the director positions cameras: Walter, *The Living Brain*, 78–79.

16   Further vistas were opened: Ibid., 90, 95.

16   "master discipline": Pressman, *Last Resort*, 36–37.

16   Walter also trained: Walter, *The Living Brain*, 99.

17   the moviegoer who: Ibid., 98.

17   effects he compared: Ibid., 188–89.

17   "fashion a mirror": Ibid., 133.

18   "We called him tortoise": W. Grey Walter, "An Imitation of Life," *Scientific American*, May 1950, 42.

18   Festival of Britain: cyberne1, "W. Grey Walter and the Festival of Britain (1951)," Cyberneticzoo.com, September 14, 2009, http://cyberneticzoo.com/cybernetic animals/w-grey-walter-and-the-festival-of-britain-1951/; Lisa Mullen, *Midcentury Gothic: The Uncanny Objects of Modernity in British Literature and Culture After the Second World War* (Manchester, UK: Manchester University Press, 2019), 99–104.

19   "In difficult situations": Walter, "An Imitation of Life."

19   Essentially he was driving: Pickering, *The Cybernetic Brain*, 67.

20   "Switching off all the circuits": Walter, *The Living Brain*, 184–85. See also 227 ff.

20   Kubie raised a series: Kubie, comment on Walter, "Where Vital Things Happen," April 17, 1959, Box 62, Grey Walter, Lawrence Kubie Papers, LOC.

20   "Your writing encourages me": Walter to Kubie, April 20, 1955; see also Walter to Kubie, May 11, 1959, Box 62, Lawrence Kubie Papers, LOC.

21   "In hypnosis": Walter, *The Living Brain*, 250–51.

21   In the case of psychosomatic disorders: Ibid., 147.

21   "The will of the subject": Ibid., 106.

22   helped inspire the principles: Ibid., 148–50; Pickering, *The Cybernetic Brain*; Hayward, "The Tortoise and the Love-Machine."

22   brain in a vat: Cathy Gere, "'Nature's Experiment': Epilepsy, Brain Localization, and the Emergence of the Cerebral Subject," in *Neurocultures: Glimpses into an Expanding Universe*, edited by Francisco Ortega and Fernando Vidal (Frankfurt am Main, Germany: Peter Lang, 2011), 244–45.

22   Hans Berger: Borck, *Brainwaves*, 33, 38.

23   Estimates are: Pressman, *Last Resort*.

23   even a lobotomized brain: Walter, *The Living Brain*, 229–30.

24   "otherwise the casualties": Ibid., 230.

24   Norbert Wiener: W. Norbert, "The Brain," in *Crossroads in Time*, edited by Groff Conklin (New York: Permabooks, 1952), 299–312.

25 "From the confusion": Walter, *The Living Brain*, 258.

25 "The physiologist, viewing ": Ibid., 259–60.

25 Many of its leading members: Peter Galison, "The Ontology of the Enemy: Norbert Wiener and the Cybernetic Vision," *Critical Inquiry* 21, no. 1 (1994): 228–66; Alan McComas, *Sherrington's Loom: An Introduction to the Science of Consciousness* (New York: Oxford University Press, 2020); Hayward, "The Tortoise and the Love-Machine," 624, n. 18.

26 remains a matter of some debate: Max Stadler, "The Neuromance of Cerebral History," in *Critical Neuroscience: A Handbook of the Social and Cultural Contexts of Neuroscience*, edited by Suparna Choudhury and Jan Slaby (London: Wiley-Blackwell, 2012), 135–58; Yvan Prkachin, "'The Sleeping Beauty of the Brain': Memory, MIT, Montreal, and the Origins of Neuroscience," *ISIS* 112, no. 1 (March 2021): 22–44.

26 Some of its acolytes: Loren R. Graham, *Science and Philosophy in the Soviet Union* (New York: Alfred A. Knopf, 1974), 340.

26 A conference held: "Meeting on Control Mechanisms in Machines, Animals, Communities," May 7, 1949, RF, RG 1.1 2.1, Series 300, 401 A, Box 15, 209, Burden 1946–1949, RAC.

26 hydrogen bomb: David Halberstam, *The Fifties* (New York: Fawcett, 1993), 92–93.

26 ambivalence of cybernetics: Galison, "The Ontology of the Enemy."

26 "The enemy may be": Wiener, *The Human Use of Human Beings*, 128.

27 he ventured the hope: Galison, "The Ontology of the Enemy," 253–54.

27 "nothing more than apprentice sorcerers": Wiener, *The Human Use of Human Beings*, 129–30.

27 "so far brain physiology": Walter, *The Living Brain*, xx–xxi.

28 Deeper knowledge of the brain: Ibid., 268.

28 If society, he wrote: Ibid., 267–68, 264.

29 "The mechanism breaks down": Ibid., 265–66.

29 The brain that emerged: Stadler, "The Neuromance of Cerebral History."

29 In 1970, Walter was involved: Walter, "My Miracle."

Clinical Tale 1: H.M.

31 "the permanent present tense": Suzanne Corkin, *Permanent Present Tense: The Man with No Memory, and What He Taught the World* (New York: Allen Lane, 2013).

32 "His brain is the closest": Mike Jay, "Argument with Myself," *London Review of Books*, May 23, 2013, 16.

32 the most uncanny of all diseases: Sigmund Freud, *Studies in Parapsychology* (New York: Collier Books, 1963), 49.

32 "sacred disease": Owsei Temkin, *The Falling Sickness: A History of Epilepsy from the Greeks to the Beginnings of Modern Neurology* (Baltimore, MD: Johns Hopkins University Press, 1945).

32 Up to the end: Luke Dittrich, *Patient H.M.: A Story of Memory, Madness and Family Secrets* (New York: Random House, 2017), 165.

33 EEG also helped remove: Yvan Prkachin, "Wired Together: The Montreal Neurological Institute and the Origins of Modern Neuroscience, 1928–1965," PhD dissertation, Harvard University, 2018, 117, 158.

33 another form of brain surgery: Elliot S. Valenstein, *Great and Desperate Cures: The Rise and Decline of Psychosurgery and Other Radical Treatments for Mental Illness* (New York: Basic Books, 1986), 157; Jack D. Pressman, *Last Resort: Psychosurgery and the Limits of Medicine* (Cambridge, UK: Cambridge University Press, 1998).

33 performing it on both: Dittrich, *Patient H.M.*, 223.

34 Institute for Living: Ibid., 153.

34 At 1973 congressional hearings: Ibid., 367–70.

34 "hazardous and invasive experiments": Cathy Gere, "The Science of Pain and Pleasure in the Shadow of the Holocaust," in *Science and Emotions After 1945: A Transatlantic Perspective*, edited by Frank Biess and Daniel M. Gross (Chicago: University of Chicago Press, 2014), 143. See also Valenstein, *Great and Desperate Cures*.

34 In particular, psychosurgery catalyzed: Pressman, *Last Resort*, 356–59; Dittrich, *Patient H.M.*, 204.

34 "castration anxiety" and "homosexual" tendencies: Dittrich, *Patient H.M.*, 209.

35 Scoville published: W. B. Scoville, "The Limbic Lobe in Man," *Journal of Neurosurgery* 1, no. 1 (1954): 64–66.

36 Hebbian synapse: Gordon M. Shepherd, *Creating Modern Neuroscience: The Revolutionary 1950s* (New York: Oxford University Press, 2010), 158–64.

36 "Here not much planning": Robert S. Morison, interview with D. Hebb, November 1, 1951, RF, 1.2, 427, 3, 33, RAC.

36 "analysis of disordered function": Brenda Milner, autobiography, in *The History of Neuroscience in Autobiography*, vol. 2, edited by Larry R. Squire (Washington, DC: Society for Neuroscience, 1998), 280.

37 "future of neuropsychology": Ibid., 285.

37 "founding text": Dittrich, *Patient H.M.*, 233.

37 a moment of central significance: Shepherd, *Creating Modern Neuroscience*, 161, 174; Eric R. Kandel, *In Search of Memory: The Emergence of a New Science of Mind* (New York: W. W. Norton, 2006).

37 "Memory was the sleeping beauty": Interview with Brenda Milner, December 2, 2011, quoted in Yvan Prkachin, "'The Sleeping Beauty of the Brain': Memory, MIT, Montreal, and the Origins of Neuroscience," ISIS 112, no. 1 (March 2021): 22–44.

38 Confronted on one test: Dittrich, *Patient H.M.*, 304.

38 "he lost his humanness": Ibid., 275, 288.

38 The story of Henry Molaison: Luke Dittrich, "The Brain That Couldn't Remember," *New York Times Magazine*, August 3, 2016, https://www.nytimes.com /2016/08/07/magazine/the-brain-that-couldnt-remember.html.

38 The disposition of Einstein's brain: Steven Levy, "Yes, I Found Einstein's Brain," *Wired*, April 17, 2015, https://www.wired.com/2015/04/yes-i-found-einsteins -brain/.

39 Julius Hallervorden: Ulf Schmidt, *Justice at Nuremberg: Leo Alexander and the Nazi Doctors' Trial* (London: Palgrave MacMillan, 2004), 272.

Chapter 2: The Brain Is an Undiscovered Country

40   the Montreal Neurological Institute: Herbert Jasper, autobiography, in *The History of Neuroscience in Autobiography*, vol. 1, edited by Larry R. Squire (Washington, DC: Society for Neuroscience, 1996), 318–47; Brenda Milner, autobiography, in *The History of Neuroscience in Autobiography*, vol. 2, edited by Larry R. Squire (Washington, DC: Society for Neuroscience, 1996), 276–305; Gordon M. Shepherd, *Creating Modern Neuroscience: The Revolutionary 1950s* (New York: Oxford University Press, 2010), 164.

41   The operating theater: Alison Winter, *Memory: Fragments of a History* (Chicago: University of Chicago Press, 2013), 80. I'm deeply grateful to Dr. Richard LeBlanc for the guided tour of the MNI and Penfield's operating theater he generously provided me.

41   stories about them: Milton Silverman, "Now They're Exploring the Brain," *Saturday Evening Post*, October 8, 1949.

42   European medical traditions: John Eccles and William Feindel, "Wilder Graves Penfield, 26 January 1891–5 April 1976," *Biographical Memoirs of Fellows of the Royal Society* 24 (1978): 471–513.

42   "undiscovered country": Wilder Penfield, *No Man Alone: A Neurosurgeon's Life* (Boston: Little, Brown and Company, 1977), 36.

42   "the master organ": Wilder Penfield, "The Interpretive Cortex," *Science* 129, no. 3365 (1959): 1719–25.

43   Penfield's patients thus became: Winter, *Memory*, 80–82.

43   "the living, thinking": Penfield, *No Man Alone*, 133.

43   "vocalize": Quoted in Silverman, "Now They're Exploring the Brain"; Penfield, *No Man Alone*, 133.

43   "It must be great fun": Eccles and Feindel, "Wilder Graves Penfield," 482.

44   "stream of consciousness": Wilder Penfield, "Studies of the Cerebral Cortex of Man: A Review and an Interpretation," in *Brain Mechanisms and Consciousness: A Symposium Organized by the Council for International Organizations of Medical Sciences*, edited by Edgar E. Adrian, Frederic Bremer, and Herbert H. Jasper (Oxford, UK: Blackwell Scientific Publications, 1954), 67–68; Wilder Penfield, "The Permanent Record of the Stream of Consciousness," in *Proceedings of the 14th International Congress of Psychology*, Montreal, June 1954, 67–68. See also Winter, *Memory*, 92.

44   "a strip of film": Penfield, "Studies of the Cerebral Cortex of Man," 295.

44   "projected on the screen": Penfield, quoted in Winter, *Memory*, 92.

44   "reminiscences": Oliver Sacks, *The Man Who Mistook His Wife for a Hat* (New York: Simon & Schuster, 1970), 137–38, 145; Oliver Sacks, *Hallucinations* (New York: Vintage, 2012), 152–53.

45   she remained somewhat skeptical: Milner, autobiography, 283.

45   It was William Scoville's: Eric R. Kandel, *In Search of Memory: The Emergence of a New Science of Mind* (New York: W. W. Norton, 2006), 127–33.

46   "bowl of porridge": Ibid., 116, 133.

46   black box: Shepherd, *Creating Modern Neuroscience*, 164.

46   a new conception: Katja Guenther, *Localization and Its Discontents: A Genealogy*

of Psychoanalysis and the Neuro Disciplines (Chicago: University of Chicago Press, 2015).

46 Penfield, who referred: Ibid., 173.

46 an information-processing system: Shepherd, *Creating Modern Neuroscience*, 228; Joseph LeDoux, *Anxious: Using the Brain to Understand and Treat Fear and Anxiety* (New York: Penguin, 2015), 26; Nicolas Langlitz, *Neuropsychedelia: The Revival of Hallucinogen Research Since the Decade of the Brain* (Berkeley: University of California Press, 2012).

47 "formalized information": Flo Conway and Jim Siegelman, *Dark Hero of the Information Age: In Search of Norbert Wiener* (New York: Basic Books, 2005), 278; see also Shepherd, *Creating Modern Neuroscience*.

47 one of the great scientific problems: Oliver Sacks, *On the Move: A Life* (New York: Vintage, 2014), 344.

47 As Ian Hacking wrote: Ian Hacking, introductory essay, in Thomas S. Kuhn, *The Structure of Scientific Revolutions*, 50th anniversary edition (Chicago: University of Chicago Press, 2012 [1962]), vii–xxxiii.

48 Penfield's ongoing: "Temporal Lobe Research Project," February 9, 1953, Wilder Penfield Digital Collection, Osler Library of the History of Medicine, McGill University, https://digital.library.mcgill.ca/penfieldfonds/fullrecord.php?ID= 9837&d=1.

48 to discover where: Kandel, *In Search of Memory*, 55, 126.

49 "An Iowan farm girl": Lawrence S. Kubie, "Some Implications for Psychoanalysis of Modern Concepts of the Organization of the Brain," *Psychoanalytic Quarterly* 22 (1953): 43.

50 He stressed the need: Ibid., 23.

50 Freud's psychic agencies: Ibid., 43, 30–32.

51 "To quote Penfield": Ibid., 44.

51 "psychoanalysts who are": Ibid., 47.

51 "It seems to this author": Ibid., 34.

52 "very few people": Kubie to Wilder Penfield, June 21, 1954, Lawrence S. Kubie Papers, LOC.

53 he wrote that: Kubie to Stanley Cobb, December 21, 1953, Lawrence S. Kubie Papers, LOC.

53 Cobb's response: Stanley Cobb to Kubie, December 31, 1953, Lawrence S. Kubie Papers, LOC.

53 Penfield's interest in mind: Guenther, *Localization and Its Discontents*, 187.

53 Later research suggested: Kandel, *In Search of Memory*, 126. But see also Winter, *Memory*, 258; Oliver Sacks, "Reminiscence," in *The Man Who Mistook his Wife for a Hat*, 132–50.

53 "seemed highly implausible": Milner, autobiography, 283–84.

54 by the end of that period: Nathan G. Hale, Jr., *The Rise and Crisis of Psychoanalysis in the United States, Freud and the Americans, 1917–1985*, vol. 2 of *Freud in America* (New York: Oxford University Press, 1995), 257; for an account of one major turf battle between the psychoanalysts and their professional rivals, see Gerald N. Grob,

The Mad Among Us: A History of the Care of America's Mentally Ill (New York: Free Press, 2011), 339–40.

54 he viewed most: Guenther, *Localization and Its Discontents*, 173–74.

55 normalized psychological breakdown: Ellen Herman, *The Romance of American Psychology: Political Culture in the Age of Experts* (Berkeley: University of California Press, 1995), 112.

55 pioneered new methods of treatment: Roy R. Grinker and John P. Spiegel, *Men Under Stress* (Philadelphia: Blakiston, 1945).

56 The release in 1952: Eli Zaretsky, *Secrets of the Soul: A Social and Cultural History of Psychoanalysis* (New York: Alfred A. Knopf, 2004).

56 "a seminal moment": David Healy, *The Creation of Psychopharmacology* (Cambridge, MA: Harvard University Press, 2002), 4.

56 broader class of psychopharmaceuticals: Between 1948 and 1958, "all prototypes of modern psychopharmaceuticals . . . were discovered"; Shepherd, *Creating Modern Neuroscience*, 206–7.

57 By the end of the twentieth century: Healy, *The Creation of Psychopharmacology*; Edward Shorter, *A History of Psychiatry: From the Era of the Asylum to the Age of Prozac* (New York: Wiley, 1998); Allen Frances, "Entrenched Reductionisms: The Bête Noire of Psychiatry," *History of Psychology* 19, no. 1 (2016): 57–59; Nikolas Rose, "Neurochemical Selves," *Society* 41, no. 1 (2003): 46–59.

57 "patient seems to be": Silverman, "Now They're Exploring the Brain."

58 Penfield's hope: Yvan Prkachin, "Wired Together: The Montreal Neurological Institute and the Origins of Modern Neuroscience, 1928–1965," PhD dissertation, Harvard University, 2018, 207–56.

58 "grand cure": Rebecca Lemov, "Brainwashing's Avatar: The Curious Career of Dr. Ewan Cameron," *Grey Room* 45 (2011): 61–87.

58 Cameron read voraciously: Ibid., 75.

58 taped loops that played endlessly: Ibid.; Anne Collins, *In the Sleep Room: The Story of the CIA Brainwashing Experiments in Canada* (Toronto: Key Porter Books, 1997), 122, 196.

59 He reported that: William Sargant, *Battle for the Mind: A Physiology of Conversion and Brainwashing* (Garden City, NY: Doubleday, 1957), xx; Ben Shephard, *A War of Nerves* (Cambridge, MA: Harvard University Press, 2001), 209.

59 "psychiatric adventurers": Winter, *Memory*, 58.

60 "the civil rights movement": Rebecca Lemov, *World as Laboratory: Experiments with Mice, Mazes, and Men* (New York: Macmillan, 2013), 218.

60 The historian of science: Cathy Gere, "'Nature's Experiment': Epilepsy, Brain Localization, and the Emergence of the Cerebral Subject," in *Neurocultures: Glimpses into an Expanding Universe*, edited by Francisco Ortega and Fernando Vidal (Frankfurt am Main, Germany: Peter Lang, 2011), 245–46.

60 "By varying the program": Hilary Putnam, *Reason, Truth and History* (New York: Cambridge University Press, 1981), 6. My thanks to Gary Wolf for drawing my attention to Putnam's thought experiment.

Clinical Tale 2: Operators and Things

63 "By the time": Barbara O'Brien, *Operators and Things: The Inner Life of a Schizophrenic* (Los Angeles: Silver Birch Press, 2011 [1958]), 52.

64 "to probe": Ibid., 74, 199, 202.

64 "Your mind is 'split'": Ibid., 23.

65 "is a wide-screen movie": Ibid., 154.

65 "The world of operators": Mike Jay, *The Influencing Machine: James Tilley Matthews and the Air Loom* (London: Strange Attractor Press, 2003), 249.

65 they had become integral: Alison Winter, *Memory: Fragments of a History* (Chicago: University of Chicago Press, 2013).

65 Lashley had offered: Karl Lashley, cited in Luke Dittrich, *Patient H.M.: A Story of Memory, Madness and Family Secrets* (New York: Random House, 2017), 185–86; see also Cornelius Borck, *Brainwaves: A Cultural History of Electroencephalography* (New York: Routledge, 2018).

66 "The unconscious has": O'Brien, *Operators and Things*, 170, 181.

66 she chose to maintain: Ibid., 7.

67 "brainless": Allen Frances, "Entrenched Reductionisms: The Bête Noire of Psychiatry," *History of Psychology* 19, no. 1 (2016): 57.

67 antipsychiatrists: Gregory Bateson, introduction, in John Perceval, *Perceval's Narrative*, edited by Gregory Bateson (London: The Hogarth Press, 1962); R. D. Laing, *The Politics of Experience* (London: Harmondsworth, 1967).

68 "Mad Scientist": O'Brien, *Operators and Things*, 69.

68 "brainless": Andrew Pickering, *The Cybernetic Brain: Sketches of Another Future* (Chicago: University of Chicago Press, 2010), 8.

68 "How very odd": O'Brien, *Operators and Things*, 36.

68 "emotional upheaval": Ibid., 146.

Clinical Tale 3: Lieutenant Dulles

71 "I beg your pardon sir": Clover Todd Dulles (hereafter CTD), journal, December 29, 1952, Allan Macy Dulles (hereafter AMD) Papers, Seeley G. Mudd Manuscript Library, Princeton University, Princeton, NJ. Permission to cite kindly granted by Joan Talley Dulles.

72 "Oh father, I've been": CTD, journal, December 20, 1952, AMD Papers.

72 "My mind is very evil now": CTD, journal, December 28, 1952, AMD Papers.

72 "Mother could you tie me": CTD, journal, December 29, 1952, AMD Papers.

73 "Brain Warfare": Allen Welsh Dulles (hereafter AWD), "Brain Warfare," May 1953, Box 61, folder 9, AWD Papers, Seeley G. Mudd Manuscript Library, Princeton University, Princeton, NJ.

74 As Penfield wrote: Wilder Penfield to AMD, January 4, 1953, AMD Papers.

74 the dark climate of that time: Peter S. Mueller Report, June 13, 1994, AMD Papers.

74 a psychological profile of Hitler: Peter Grose, *Gentleman Spy: The Life of Allen Dulles* (New York: Houghton Mifflin Harcourt, 1994), 245.

74   His inability to form: Joan Talley Dulles, personal communication, December 31, 2015.

74   "Make a note": CTD, journal, December 29, 1952, AMD Papers.

75   investigated by CIA personnel: Stephen Kinzer, *Poisoner in Chief: Sidney Gottlieb and the CIA Search for Mind Control* (New York: Holt, 2019), 40.

76   "Man himself": Ladislas Farago, ed., *German Psychological Warfare* (New York: Committee on National Morale, 1942).

76   "the Dulles era": Grose, *Gentleman Spy*, 338.

76   Dulles's carefully cultivated image: Tim Weiner, *Legacy of Ashes: The History of the CIA* (New York: Anchor Books, 2007), 107.

76   in a speech: Dulles, "Brain Warfare."

76   a fundamental contradiction: Weiner, *Legacy of Ashes*, 28, 578–59; Timothy Melley, *The Covert Sphere: Secrecy, Fiction, and the National Security State* (Ithaca, NY: Cornell University Press, 2012).

77   "Manhattan Project": Aldous Huxley, new preface to *Brave New World* (New York: Penguin, 1946).

78   a kind of clinical case: David Riesman, Nathan Glazer, and Reuel Denney, *The Lonely Crowd* (New Haven, CT: Yale University Press, 1950); Vance Packard, *The Hidden Persuaders* (New York: David McKay Company, 1957).

Chapter 3: An Experiment with Mankind

79   "In Red China": Leon Freedom, "Corticovisceral Psychiatry," in Background to *Brainwashing in Red China*, Edward Hunter papers, 6–7.

79   obscure Soviet film: Edward Hunter, *Brainwashing: The Story of Men Who Defied It* (New York: Farrar Strauss and Cudahy, 1956).

80   Hunter had first described: Edward Hunter, *Brain-Washing in Red China: The Calculated Destruction of Men's Minds* (New York: Vanguard, 1951).

81   "continuous war": Paul Myron Anthony Linebarger, *Psychological Warfare* (New York: Duell, Sloan and Pierce, 1948).

81   questions of morale, belief, and loyalty: Ellen Herman, *The Romance of American Psychology: Political Culture in the Age of Experts* (Berkeley: University of California Press, 1995), 30.

81   "Total war inevitably made man": Ladislas Farago, ed., *German Psychological Warfare* (New York: National Committee of Morale, 1942), 47.

81   Such a view was echoed: Ron Robin, *The Making of the Cold War Enemy: Culture and Politics in the Military-Intellectual Complex* (Princeton, NJ: Princeton University Press, 2001).

81   a form of warfare: Herman, *The Romance of American Psychology*, 124.

81   Recent revelations: Allen W. Dulles, "Brain Warfare—Russia's Secret Weapon," *U.S. News & World Report*, May 8, 1953, 54.

82   Here we encounter: Cathy Gere, "'Nature's Experiment': Epilepsy, Brain Localization, and the Emergence of the Cerebral Subject," in *Neurocultures: Glimpses into an Expanding Universe*, edited by Francisco Ortega and Fernando Vidal (Frankfurt am Main, Germany: Peter Lang, 2011), 235–48.

82   "The Communists have recently": Dulles, "Brain Warfare," 58.

83   a 1951 JCS document: Robin, *The Making of the Cold War Enemy*, 42.

83   thirty-eight US Air Force POWs: Abbott Gleason, *Totalitarianism: The Inner History of the Cold War* (New York: Oxford University Press, 1995), 99.

84   first direct confrontation: Robin, *The Making of the Cold War Enemy*, 8; John Lewis Gaddis, *The Cold War: A New History* (New York: Penguin, 1995); Bruce Cumings, *The Korean War: A History* (New York: Modern Library Chronicles, 2010).

84   "garrison state": Lasswell, quoted in Jonathan D. Moreno, *Mind Wars: Brain Research and National Defense* (New York: Dana Press, 2006), 21.

84   Korea became the first test: Gaddis, *The Cold War*, 91–92.

84   Though social scientific works: Hannah Arendt, *The Origins of Totalitarianism* (New York: Harcourt Brace Jovanovich, 1951).

84   "When I try to picture": George Kennan, 1953, quoted in Susan L. Carruthers, *Cold War Captives: Imprisonment, Escape, and Brainwashing* (Berkeley: University of California Press, 2009), 13–14.

84   "brainwashing": Timothy Melley, *The Covert Sphere: Secrecy, Fiction, and the National Security State* (Ithaca, NY: Cornell University Press, 2012).

85   "police action": David Halberstam, *The Fifties* (New York: Fawcett, 1993), 70.

85   According to the historian Bruce Cumings: Cumings, *The Korean War*, 174–77.

85   In their efforts: Gaddis, *The Cold War*, 90, 105.

85   The problem: Halberstam, *The Fifties*, 67.

86   "came along and saved us": Acheson, quoted in Cumings, *The Korean War*, 210.

86   The invasion of South Korea: Cumings, *The Korean War*, 205.

86   "We who live": Ibid., 99.

86   shrouded in mystery: Ibid., 79.

86   American security: Gaddis, *The Cold War*, 90.

87   "ghastly experiment": Arendt, *The Origins of Totalitarianism*, 438, 455.

88   Surprisingly little is known: Edward Hunter, "Communist Psychological Warfare (Brainwashing)," testimony before the Committee on Un-American Activities, House of Representatives, March 13, 1958, https://archive.org/stream/communist psychol1958unit/communistpsychol1958unit_djvu.txt.

89   "plausible but fictive chain": Darius Rejali, *Torture and Democracy* (Princeton, NJ: Princeton University Press, 2007), 71.

89   "folklore of industrial man": Marshall McLuhan, *The Mechanical Bride: Folklore of Industrial Man* (Boston: Beacon Press, 1951).

89   "brain operation": Hunter, *Brain-Washing in Red China*, 301.

89   It opened with a brief account: Ibid., 5–10.

92   He then went on: Ibid., 11–12.

93   For Arendt, the essence: Arendt, *The Origins of Totalitarianism*; Gleason, *Totalitarianism*, 113; Andreas Killen, "*Homo pavlovius*: Cinema, Conditioning, and the Cold War," *Grey Room* 45 (2011): 42–59.

93   "*The formula for man*": Robert Tucker, "Stalin and the Uses of Psychology," RAND Corporation, March 10, 1955, https://www.rand.org/content/dam/rand/pubs /research_memoranda/2006/RM1441.pdf, 25.

94   "cortico-visceral physiology": Ibid., 40.

94   second signal system: Ibid., 65.

95   "For a time": Gleason, *Totalitarianism*, 101.

95   "one of the least motivated": Ben Shephard, *A War of Nerves: Soldiers and Psychiatrists in the Twentieth Century* (Cambridge, MA: Harvard University Press, 2001), 341.

95   "as hard for ordinary": Samuel Dean to Edward Hunter, July 18, 1953, Corr. 1930–54, Hunter Papers, Wisconsin Historical Society.

95   Indeed, official reports: Robin, *The Making of the Cold War Enemy*, 164.

95   "dramatic film": Hunter to Westbrook, April 3, 1952, Corr. 1930–54, Hunter Papers.

96   most Americans: Hunter, *Brainwashing*, 5.

96   Yet the phenomenon: Killen, "*Homo pavlovius.*"

96   "peculiar feeling": Hunter, *Brainwashing*, 18.

97   "The twinge was involuntary": Ibid., 23.

98   Pavlov was commissioned: Ibid., 39–43.

99   "As I was taking notes": Hunter, "Communist Psychological Warfare (Brainwashing)," 13.

99   "Red POW camps": Hunter, *Brainwashing*, 253.

100   a chapter based on extended conversations: Leon Freedom, Background to *Brainwashing in Red China*, n.d., Hunter Papers.

100   In essence, this was: Hunter, *Brainwashing*, 249, 261.

100   "exactly as the hysterical": Leon Freedom, "Menace to the Free World," in Background to *Brainwashing in Red China*, Hunter Papers.

101   Freedom cited the speculative futures: Robin, *The Making of the Cold War Enemy*, 164.

101   "application to human beings": "Brainwashing: The Communist Experiment with Mankind," April 19, 1955, NSC Papers, Dwight D. Eisenhower Presidential Library, Abilene, KS. See also David Seed, *Brainwashing: The Fictions of Mind Control* (Kent, OH: Kent State University Press, 2004), 53–56.

102   Attached to the document: Evaluation of booklet "Brainwashing, A Synthesis of the Russian Textbook on Psycho-politics," March 20, 1956, NSC Papers, Eisenhower Presidential Library. See also Seed, *Brainwashing*, 41–44.

102   the CIA fell under the spell: Stephen Kinzer, *Poisoner in Chief: Sidney Gottlieb and the CIA Search for Mind Control* (New York: Holt, 2019), 54; Andreas Killen and Stefan Andriopoulos, "Editors' Introduction: On Brainwashing: Mind Control, Media, and Warfare," *Grey Room* 45 (2011): 7–17.

102   "As for the work": Evaluation of booklet "Brainwashing, A Synthesis of the Russian Textbook on Psycho-politics."

104   "peculiar modern conceit": Rejali, *Torture and Democracy*, 71. See also Rebecca Lemov, "Brainwashing's Avatar: The Curious Case of Dr. Ewen Cameron," in *Grey Room* 45 (2011): 60–87. For a skeptical contemporary account of the relevance of Pavlov, see Raymond Augustine Bauer, *The New Man in Soviet Psychology* (Cambridge, MA: Harvard University Press, 1952).

104   That constellation created conditions: Rebecca Lemov, *World as Laboratory:*

Experiments with Mice, Mazes, and Men (New York: Macmillan, 2013); Melley, *The Covert Sphere*.

104   "No mental theory": Edward Hunter to Virginia Freedom, December 12, 1955, Corr. 1954–60, Hunter Papers.

Clinical Tale 4: Hayes's Hallucination

107   "The first thing": Richard Condon, *The Manchurian Candidate* (New York: Four Walls Eight Windows, 2003 [1959]), 32.

107   "It is man's consciousness": "Brainwashing: The Communist Experiment with Mankind," April 19, 1955, NSC Papers, Dwight D. Eisenhower Presidential Library, Abilene, KS.

108   The accusations against Schwable and Batchelor: See, e.g., *United States v. Claude J. Batchelor*, 466; *Court Martial Reports: Holdings and Decisions of the Judge Advocates General and United Court of Military Appeals* 19 (1955).

108   Schwable somewhat disturbingly: Elie Abel, "Schwable Tells of P.O.W. Ordeal," *New York Times*, March 12, 1954, https://timesmachine.nytimes.com/times machine/1954/03/12/110067999.html.

108   "mental disorder": *United States v. Batchelor*, 489. See also Brian D. McKnight, *We Fight for Peace: Twenty-Three American Soldiers, Prisoners of War, and Turncoats in the Korean War* (Kent, OH: Kent State University Press, 2014), 188–92; Susan L. Carruthers, *Cold War Captives: Imprisonment, Escape, and Brainwashing* (Berkeley: University of California Press, 2009), 296, n. 130.

108   Edward Hunter's correspondence with Freedom: Leon Freedom to Edward Hunter, October 6, 1954, Corr. 1954–60, Hunter Papers.

108   They are echoed: Westbrook to Mrs. Batchelor, October 6, 1954, Corr. 1954–60, Hunter Papers. See also *United States v. Batchelor*, 454.

109   legitimacy of the brainwashing defense: McKnight, *We Fight for Peace*, 187–88; Carruthers, *Cold War Captives*, 293, n. 83; 295, n. 104; 296, n. 131.

109   literary portrayals: Westbrook to Hunter, April 17, 1959, Corr. 1954–60, Hunter Papers; Virginia Pasley, *22 Stayed* (New York: W. H. Allen, 1955), 211.

109   "We admitted no private sphere": Arthur Koestler, *Darkness at Noon* (New York: The MacMillan Company, 1940), 80.

109   Rubashov passes the time: Ibid., 12.

110   "the brains": Ibid., 48.

110   John B. Hayes: Edward Hunter, *Brainwashing: The Story of Men Who Defied It* (New York: Farrar Strauss and Cudahy, 1956), 64–88.

110   Hayes had worked: Ibid., 76–78.

112   "If truth can linger": Ibid., 88.

Chapter 4: The "Prisoner's Cinema": Interrogation in the Age of the Explorable Brain

114   "He was rolling": George Orwell, *1984* (New York: Penguin, 2003 [1949]), 250.

114   "There was a certain": Ibid., 254.

114   As Hannah Arendt observed: Hannah Arendt, *The Origins of Totalitarianism* (New York: Harcourt Brace Jovanovich, 1951), 434.

114   Watson's experiments with fear conditioning: Rebecca Lemov, *World as Laboratory: Experiments with Mice, Mazes, and Men* (New York: Macmillan, 2013), 41–42.

114   As Darius Rejali has observed: Darius Rejali, *Torture and Democracy* (Princeton, NJ: Princeton University Press, 2007), 458.

115   "We make the brain perfect": Orwell, *1984*, 263.

115   "*1984* is closer than we think": Leon Freedom, "Cortico-Visceral Psychiatry," n.d., Hunter Papers.

116   the film *POW*: Susan L. Carruthers, *Cold War Captives: Imprisonment, Escape, and Brainwashing* (Berkeley: University of California Press, 2009), 198; J. Hoberman, *An Army of Phantoms: American Movies and the Making of the Cold War* (New York: The New Press, 2011), 306–7.

116   "the world's favorite conspiracy theory": Dominic Streatfeild, *Brainwash: The Secret History of Mind Control* (New York: Picador, 2008), 342.

116   One of many reminders: Jane Mayer, *The Dark Side: The Inside Story of How the War on Terror Turned into a War on American Ideals* (New York: Anchor, 2009); Alfred McCoy, *A Question of Torture: CIA Interrogation, from the Cold War to the War on Terror* (New York: Holt, 2006); Naomi Klein, *The Shock Doctrine: The Rise of Disaster Capitalism* (New York: Picador, 2008), 29–58; Lemov, *World as Laboratory*; Rejali, *Torture and Democracy*.

116   the 2015 report: "Report of the Independent Reviewer and Related Materials," American Psychological Association, 2015, https://www.apa.org/independent-review, 30.

117   one of the comforting illusions: Rejali, *Torture and Democracy*, 458.

117   "strategic fiction": Timothy Melley, *The Covert Sphere: Secrecy, Fiction, and the National Security State* (Ithaca, NY: Cornell University Press, 2012), vii–viii, 218–21.

117   "clarify the differences": John C. Lilly, opening remarks, in *Methods of Forcible Indoctrination: Observations and Interviews*, Symposium no. 4 (New York: Group for the Advancement of Psychiatry, July 1957), 234.

118   "Manhattan Project of the mind": McCoy, *A Question of Torture*, 7.

118   science of the self: John Marks, *The Search for the "Manchurian Candidate": The CIA and Mind Control* (New York: W. W. Norton & Co., 1979); Lemov, *World as Laboratory*.

118   other scholars have challenged: Rejali, *Torture and Democracy*; Lemov, *World as Laboratory*, 201.

118   As the author John Marks put it: Marks, *The Search for the "Manchurian Candidate,"* 174, 153.

119   the CIA's LSD research: Martin Lee and Bruce Shlain, *Acid Dreams: The CIA, LSD, and the Sixties Rebellion* (New York: Grove Press, 1985).

119   "laboratory and clinical investigations": "Communist Control Techniques," National Security Archives, April 2, 1956, https://nsarchive2.gwu.edu/torturingdemocracy/documents/19560400.pdf, 3–4. See also Lawrence E. Hinkle, Jr., and Harold G.

Wolff, "Communist Interrogation and Indoctrination of 'Enemies of the State': Analysis of Methods Used by the Communist State Police (A Special Report)," CIA report, reprinted from *AMA Archives of Neurology and Psychiatry* 76, no. 2 (1956): 115–74.

119   "situation of frustration:" Hinkle and Wolff, "Communist Interrogation and Indoctrination of 'Enemies of the State,'" 112.

119   Eventually the prisoner: Ibid., 22–23, 31, 51, 63.

120   the alteration of belief: Ibid., 110–12.

120   "nevertheless the interrogation": Ibid., 115.

120   A recent biographical sketch: Stephen D. Silberstein, Richard B. Lipton, and David W. Dodick, eds., *Wolff's Headache and Other Head Pain*, 8th ed. (New York: Oxford University Press, 2007), 3.

121   "wiped almost clean": William Sargant, *Battle for the Mind: A Physiology of Conversion and Brainwashing* (London: Doubleday, 1957), 14.

122   human centrifuge: Lemov, *World as Laboratory*, 203.

122   "In an effort": Ibid., 208. See also Marks, *The Search for the "Manchurian Candidate*," 158.

122   "the human personality": SIHE Annual Report, 1957, 8–9, Box 6, folder 15, Harold Wolff Papers.

122   "age of extremes": Eric Hobsbawm, *The Age of Extremes: A History of the World, 1914–1991* (New York: Vintage, 1994).

123   "master organ": Harold Wolff, "Has Disease Meaning?," Box 9, folder 19, Harold Wolff Papers.

123   "Where any of these studies": "Effects of Chemical Agents on Bodily Functions, Mentation, Attitude, etc.," Box 6, folder 15, Harold Wolff Papers.

124   The document enumerated: "Factors Affecting Behavior, Mentation, Attitude, etc.," Box 6, folder 15, Harold Wolff Papers.

124   been produced by the CIA: Marks, *The Search for the "Manchurian Candidate*"; Stephen Kinzer, *The Poisoner in Chief: Sidney Gottlieb and the CIA Search for Mind Control* (New York: Henry Holt, 2019).

124   "solidly academic pedigree": Maarten Derksen, *Histories of Human Engineering: Tact and Technology* (New York: Cambridge University Press, 2017), 164.

125   "hypnotic crime": Stefan Andriopoulos, "The Sleeper Effect: Hypnotism, Mind Control, Terrorism," *Grey Room* 45 (2011): 88–105.

125   so-called Mindszenty look: Anne Collins, *In the Sleep Room: The Story of the CIA Brainwashing Experiments in Canada* (Toronto: Key Porter Books, 1988), 25; François Dagognet, "Towards a Bio-psychiatry," in *Incorporations*, edited by Jonathan Crary and Sanford Kwinter (New York: Zone Books, 1992), 519.

126   In his still unsurpassed account: Marks, *The Search for the "Manchurian Candidate*," 21.

127   "The process of trance-induction": Martin T. Orne, "Hypnotically Induced Hallucinations," in *Hallucinations*, edited by L. J. West (New York: Grune & Stratton, 1962), 219.

128   a list of fictional sources: SIHE Annual Report, bibliography, 1957, Box 6, folder

15, Harold Wolff Papers; David Seed, *Brainwashing: The Fictions of Mind Control* (Kent, OH: Kent State University Press, 2004), 261; Andreas Killen and Stefan Andriopoulos, "Editors' Introduction: On Brainwashing: Mind Control, Media, and Warfare," *Grey Room* 45 (2011): 7–17.

129 Orwell's depiction of the role: Donald Hebb, "The Psychology of Hocus-Pocus," Inaugural Lecture, January 12, 1950, File 01.70-0001, Hebb Fonds, McGill University Archives, Montreal.

129 "the great man of psychology": Donald Hebb, "Donald O. Hebb (1904–1985)," in *A History of Psychology in Autobiography*, vol. 7, edited by Gardner Lindzey (New York: W. H. Freeman, 1980), 288.

130 "second most important book": D. O. Hebb, *The Organization of Behavior: A Neuropsychological Theory* (New York: John Wiley & Sons, 1949). See also Gordon M. Shepherd, *Creating Modern Neuroscience: The Revolutionary 1950s* (New York: Oxford University Press, 2010), 162–64, 174.

130 a false conception: Hebb, "The Psychology of Hocus-Pocus," 10–11.

131 Hebb noted: Ibid., 14–15.

131 "The last ten years": Stanley Cobb, opening remarks, in *Sensory Deprivation: A Symposium Held at Harvard Medical School*, edited by Philip Solomon et al. (Cambridge, MA: Harvard University Press, 1961), xvii.

132 "The intactness of one's personality": Donald Hebb, "Studies of the Thinking Process," May 1, 1954, RF, RG 1.2 427, 4, 33, RAC.

132 "inner-directed personality": Robert S. Morison, diary, January 25, 1954, RF, RG 12, RAC.

133 *Alice's Adventures in Wonderland*: Donald Hebb, "The Problem of Consciousness and Introspection," in *Brain Mechanisms and Consciousness: A Symposium Organized by the Council for International Organizations of Medical Sciences*, edited by Edgar E. Adrian, Frederic Bremer, and Herbert H. Jasper (Oxford, UK: Blackwell Scientific Publications, 1954), 415–17.

133 But the onset of hallucinations: Donald O. Hebb, Annual Report, January 27, 1955, RF, 1.2, 427, 4, 34, RAC.

133 "can shake him": D. O. Hebb, "The Motivating Effects of Exteroceptive Stimulation," *American Psychologist* 13, no. 3 (1958): 111.

134 "implanting" ideas: Collins, *In The Sleep Room*, 49; Charlie Williams, Sarah Marks, and Daniel Pick, "Can Isolation Lead to Manipulation?," Wellcome Collection, September 27, 2018, https://wellcomecollection.org/articles/W1bwkyYAACUAqy10.

134 overrode their habitual skepticism: Robert S. Morison, diary, October 24, 1952, RF, RG 12, 336, RAC.

135 "There is nothing mysterious": *KUBARK* July 1963, https://documents.theblack vault.com/documents/terrorism/kubarkinterrogationmanual.pdf, 1.

135 They took as their starting point: Ibid., 52, 65.

136 cacophony of unrelated: Ibid., 76.

136 "projects the contents": Ibid., 88.

136 A notable feature: Ibid., 87–90.

136 the dynamics of the relationship: Lemov, *World as Laboratory*, 206.

137   "past horrifying ordeal": E. Bliss and K. Clark, "Visual Hallucination," in *Halluci-nations*, edited by L. J. West (New York: Grune & Stratton, 1962), 101.

137   "prisoner's cinema": Ronald K. Siegel, *Fire in the Brain: Clinical Tales of Hallucination* (New York: Dutton, 1992); Oliver Sacks, "The Prisoner's Cinema: Sensory Depri-vation," in *Hallucinations* (New York: Vintage, 2012), 34–44.

137   "theory of the imprisoned knower": D. O. Hebb, "Intelligence, Brain Function and the Theory of Mind," *Brain* 82, no. 2 (1959): 271 ff.

137   In subsequent writings: Donald Hebb, *Essays on Mind* (New York: Psychology Press, 1980), 35–37.

138   "choice of sensory patterns": Robert Morison, "General Discussion," in Adrian et al., *Brain Mechanisms and Consciousness*, 494.

138   "Your awareness of yourself": D. O. Hebb, "The American Revolution," *American Psychologist* 15, no. 12 (1960): 741.

138   "improbable tricks": D. O. Hebb, "The Mind's Eye," *Psychology Today*, May 1969, 55.

138   the earlier noted case: Siegel, *Fire in the Brain*, 217–19. See also Ronald K. Sie-gel, "Hostage Hallucinations: Visual Imagery Induced by Isolation and Life-Threatening Stress," *Journal of Nervous and Mental Disease* 172, no. 5 (1984): 269.

139   Oliver Sacks, drawing parallels: Sacks, "The Prisoner's Cinema," 196.

139   "neurological wonderland": Maitland Baldwin, "Hallucinations in Neurologic Syn-dromes," in West, *Hallucinations*, 77–85.

140   Was it, in fact, a kind of hallucination: Orne, "Hypnotically Induced Hallucina-tions."

140   "far in advance": Marks, *The Search for the "Manchurian Candidate,"* 157.

140   "helped liberate": Ibid., 173–74; Rebecca Lemov, "Brainwashing's Avatar: The Cu-rious Career of Dr. Ewen Cameron," *Grey Room* 45 (2011), 69.

140   birth of cognitive neuroscience: Lemov, "Brainwashing's Avatar," 69; Flo Conway and Jim Siegelman, *Dark Hero of the Information Age: In Search of Norbert Wiener* (New York: Basic Books, 2005), 279–80; Aldous Huxley, *The Doors of Perception* (London: Panther, 1954).

140   the SIHE provided: SIHE Annual Report, 1957, 23–24; Marks, *The Search for the "Manchurian Candidate,"* 146.

141   plagued by blind spots: Cathy Gere, "The Science of Pain and Pleasure in the Shadow of the Holocaust," in *Science and Emotions After 1945: A Transatlantic Per-spective*, edited by Frank Biess and Daniel M. Gross (Chicago: University of Chicago Press, 2014), 139–56; Jonathan D. Moreno, *Mind Wars: Brain Research and National Defense* (New York: Dana Press, 2006).

141   worshipped and adored him: Lemov, "Brainwashing's Avatar," 82; Collins, *In the Sleep Room*.

141   "You become very small": Marks, *The Search for the "Manchurian Candidate,"* 148.

142   Apparent exceptions: Ibid.

142   secrecy requirements: Donald Hebb to John Lilly, May 19, 1956; Lilly to Hebb, June 8, 1956, Box 29, John C. Lilly Papers, Special Collections & University Ar-chives, Stanford University, Stanford, CA.

142   "The work we have done": Donald Hebb, preface, in *Sensory Deprivation: A*

Symposium Held at Harvard Medical School, edited by Philip Solomon et al. (Cambridge, MA: Harvard University Press, 1961), 6.

142 The degree to which: See, for instance, Donald Hebb to Donald P. Dietrich, May 15, 1959, 2364.12.1-0077-72, Hebb Fonds; Marks, *The Search for the "Manchurian Candidate"*; McCoy, *A Question of Torture*; Richard E. Brown, "Alfred McCoy, Hebb, the CIA and Torture," *Journal of the History of the Behavioral Sciences* 43, no. 2 (2007), 205–13. See also the collection of archival documents at Naomi Klein's website: https://tsd.naomiklein.org/shock-doctrine/resources/chapter-resources.html.

143 "Should a Manhattan Project": John C. Lilly, "I Iuman Brain Research and Ethics," ca. 1960, 5A1, Box 5, John C. Lilly Papers.

Clinical Tale 5: Lajos Ruff

145 to deliver testimony: Senate Subcommittee on Scope of Soviet Activity in the United States. Hearing Before the Subcommittee to Investigate the Administration of the Internal Security Act and Other Internal Security Laws of the Committee on the Judiciary December 19, 1956 (Washington, DC: US Government Printing Office, 1957), 3242–3300.

147 "Bolshevism's laboratory techniques": Hannah Arendt, *The Origins of Totalitarianism* (New York: Harcourt Brace Jovanovich, 1951), 352–53: Thanks to Andras Kisery for finding the "émigré pulp fiction" quote.

147 "the most frightening": Lajos Ruff, *The Brain-Washing Machine* (London: Robert Hale, 1959 [1958]), 71–72.

148 "I even desired": Ibid., 78.

148 "It was impossible": Ibid., 86.

149 "the burden of free choice": Ibid., 90.

149 "the longest treatment": Ibid., 91.

149 accounts of the torture: Darius Rejali, *Torture and Democracy* (Princeton, NJ: Princeton University Press, 2007), 390.

150 "It is now possible": "The Struggle for the Control of the New Techniques of Conditioning," translated by Reuben Keehan, *Internationale Situationniste*, June 1958, https://www.cddc.vt.edu/sionline/si/struggle.html.

150 "It should be understood": Ibid.

Clinical Tale 6: Paul Linebarger/Cordwainer Smith

153 an article on advances in computing: "The Thinking Machine," *Time*, January 23, 1950, http://content.time.com/time/subscriber/article/0,33009,858601-5,00.html.

153 Donald Hebb mused about: Donald Hebb, "The Problem of Consciousness and Introspection," in *Brain Mechanisms and Consciousness: A Symposium Organized by the Council for International Organizations of Medical Sciences*, edited by Edgar E. Adrian, Frederic Bremer, and Herbert H. Jasper (Oxford, UK: Blackwell Scientific Publications, 1954), 411.

154 could coexist: David Seed, *Brainwashing: The Fictions of Mind Control* (Kent, OH: Kent State University Press, 2004), 53–56. I'm indebted to Seed's book for drawing my attention to Linebarger's fascinating story.

154 long considered the standard text: Paul Linebarger, *Psychological Warfare* (New York: Duell, Sloane and Pearce, 1948, revised 1954).

154 one scholar has suggested: Marcia Holmes, "Edward Hunter and the Origins of Brainwashing," Hidden Persuaders, May 26, 2017, http://www7.bbk.ac.uk/hidden persuaders/blog/hunter-origins-of-brainwashing/.

154 In a speech: Paul M. A. Linebarger, "Psychiatric Warfare," December 15, 1950, 12–13, Paul M. A. Linebarger Papers, Hoover Institution Library & Archives, Stanford, CA.

155 "black propaganda operations": Joseph Smith, *Portrait of a Cold Warrior: Second Thoughts of a Top CIA Agent* (New York: Putnam, 1976), 86–89.

156 allegedly written by: "Brainwashing: The Communist Experiment with Mankind," April 19, 1955, NSC Papers, Dwight D. Eisenhower Presidential Library, Abilene, KS; on the authorship of these texts, see Seed, *Brainwashing*, 41–45.

156 some people believing: Benjamin Mendel (Member of Internal Security Subcommittee) to J. B. Kenworthy, September 16, 1963, RG 46, Edward Hunter, National Archives, Washington, DC.

156 "The suggestions and words": "Brainwashing: The Communist Experiment with Mankind."

156 As Linebarger explained: Paul M. A. Linebarger, "Ethical Dianetics: The Lay Psychotherapy of Mutual Aid," c. 1952, Paul M. A. Linebarger Papers.

157 new science of the human mind: Jon Atack, *A Piece of Blue Sky: Scientology, Dianetics and L. Ron Hubbard Exposed* (New York: Lyle Stuart, 1990), 105.

157 As the philosopher S. I. Hayakawa: S. I. Hayakawa, "From Science-Fiction to Fiction-Science," in *ETC: A Review of General Semantics* 8, no. 4 (1951), 280–93.

157 In an unsent letter: John Campbell to Grey Walter, May 1, 1953, cited in Alec Nevala-Lee, *Astounding: John W. Campbell, Isaac Asimov, Robert A. Heinlein, L. Ron Hubbard, and the Golden Age of Science Fiction* (New York: Dey Street Books, 2019), 18–19. My thanks to Alec Nevala-Lee for sharing a copy of this letter with me.

158 warned his readers: W. Grey Walter, *The Living Brain* (London: Gerald Duckworth, 1953), 264, 251.

159 "handled like a marionette": L. Ron Hubbard, *Dianetics: The Modern Science of Mental Health* (New York: Hermitage House, 1950), xiii.

159 a vast amnesia narrative: Lawrence Wright, *Going Clear: Scientology, Hollywood, and the Prison of Belief* (New York: Vintage, 2013), 293 ff.

159 as he alleged: Atack, *A Piece of Blue Sky*, 120, 140.

159 Eventually the mass media: Wright, *Going Clear*, 146.

159 "I write this book": Linebarger, "Ethical Dianetics," 26–27.

160 "not a closed cult": Ibid., 97.

161 "Scanners Live in Vain": Seed, *Brainwashing*, 53–56; Ted Gioia, "Remembering Cordwainer Smith: Full-Time Sci-Fi Author, Part-Time Earthling," *Atlantic*, March 26, 2013.

161 "Kirk Allen": Alan C. Elms, "Behind the Jet-Propelled Couch: Cordwainer Smith and Kirk Allen," *New York Review of Science Fiction*, May 2002.

161 "He identified with": Robert Lindner, *The Fifty-Minute Hour* (London: Secker and Warburg, 1954).

161 In an essay on cybernetics: Hannah Arendt, "Cybernetics," 1964, Speeches and Writings File, 1923–1975, Essays and Lectures, Hannah Arendt Papers, LOC.

162 the confusion: Thomas Powers, *The Man Who Kept the Secrets: Richard Helms and the CIA* (New York: Knopf, 1979), 363, 381.

Chapter 5: Manchurian Candidates

163 "the most important": Aldous Huxley, new foreword to *Brave New World* (London: Chatto and Windus, 1946 [1932]), 12–13.

164 Twelve years later: Aldous Huxley, *Brave New World Revisited* (New York: Harper & Row, 1958).

164 "A passive solitary child": W. Grey Walter, *The Living Brain* (London: Gerald Duckworth, 1953), 267–68.

164 "young nervous systems": Fredric Wertham, *Seduction of the Innocent* (New York: Rinehart, 1954).

164 "Some part of my research": Wertham's congressional testimony, quoted in David Hajdu, *The Ten-Cent Plague: The Great Comic-Book Scare and How It Changed America* (New York: Picador, 2009), 263.

165 "The electroencephalograms marked": Curt Siodmak, *Donovan's Brain* (New York: Knopf, 1942), 4.

167 "get inside": Marshall McLuhan, *The Mechanical Bride: Folklore of Industrial Man* (Boston: Beacon Press, 1951), v.

167 "The ad men": Ibid., 97.

167 "take possession": Quoted in Joost A. M. Meerloo, *The Rape of the Mind: The Psychology of Thought Control, Menticide, and Brainwashing* (New York: World Pub. Comp., 1956), 11.

167 "are always trying": McLuhan, *The Mechanical Bride*, 97.

167 seminar on culture and communication: Flo Conway and Jim Siegelman, *Dark Hero of the Information Age: In Search of Norbert Wiener* (New York: Basic Books, 2005), 277.

168 "A single mechanical brain": McLuhan, *The Mechanical Bride*, 31.

168 "The misleading effect": Ibid., 93.

168 "The only functions left": Ibid., 131.

169 Packard offered: Vance Packard, *The Hidden Persuaders* (New York: Ig Publishing, 2007 [1957]), 32.

169 "Orwellian configurations": Ibid., 214.

169 "an unseen dictatorship": Ibid., 171.

169 "getting the product story": Ibid., 46–47.

169 entered into a trance state: Ibid., 113–14.

169 He also reported on experiments: Ibid., 62.

170 "Planes, missiles, and machine tools": Ibid., 219–20. Quoted material is from Curtiss Schafer in *Time* magazine.

170   misreaders: For Donald Hebb's skeptical views of Aldous Huxley, see Donald Hebb to Robert Morison, 44-0065, 44-0068, Hebb Fonds.

171   He wrote about those experiences: Aldous Huxley, *The Doors of Perception* (New York: Panther, 1954); Huxley, *Heaven and Hell* (New York: Panther, 1956).

171   he called "Mind-at-Large": Nicolas Langlitz, *Neuropsychedelia: The Revival of Hallucinogen Research Since the Decade of the Brain* (Berkeley: University of California Press, 2012), 127–28, 155–57.

172   Huxley mused: Huxley, *Heaven and Hell*, 112–15.

172   "downward" transcendence: Aldous Huxley, *The Devils of Loudun* (New York: Carroll and Graf Publishers, 1952), 320.

172   the account of Pavlov: William Sargant, *Battle for the Mind: A Physiology of Conversion and Brainwashing* (New York: Doubleday, 1957).

172   confess to literally anything: Huxley, *Brave New World Revisited*, 61–65.

173   would eventually give way: Ibid., 63, 67.

173   behaviorism, he wrote: Ibid., 103–4; see also Huxley, *The Doors of Perception*, 13.

173   had delivered a veritable cornucopia: Huxley, *Brave New World Revisited*, 77, 98.

173   a letter he wrote: Aldous Huxley, "A Letter to George Orwell," October 21, 1949, in *Brave New World Revisited*, Appendix, 8–10.

174   "that mind and brain": J.A.C. Brown, *Techniques of Persuasion: From Propaganda to Brainwashing* (New York: Penguin, 1963), 292.

174   "The will of the subject": Walter, *The Living Brain*, 106, quoted in Brown, *Techniques of Persuasion*, 305.

175   a vernacular understanding: Charles R. Acland, *Swift Viewing: The Popular Life of Subliminal Influence* (Durham NC: Duke University Press, 2012), 29.

176   warned those prone to seizures: Jimena Canales, "'A Number of Scenes in a Badly Cut Film': Observation in the Age of Strobe," in *Histories of Scientific Observation*, edited by Lorraine Daston and Elizabeth Lunbeck (Chicago: University of Chicago Press, 2011): 245.

177   visual perception functioned: Walter, *The Living Brain*, 78–81.

177   the dangers of the new medium: Alan Nadel, "Television: Cold War Television and the Technology of Brainwashing," in *American Cold War Culture*, edited by Douglas Field (Edinburgh, UK: Edinburgh University Press, 2005), 146–63.

177   From 1950 to 1954: Theodore White, *The Making of the President 1960* (New York: Atheneum, 1961), 279; David Halberstam, *The Fifties* (New York: Fawcett, 1993).

177   "in close proximity": Jonathan Crary, *24/7: Late Capitalism and the End of Sleep* (London and New York: Verso, 2014), 80.

177   The journalist Theodore H. White: White, *The Making of the President 1960*, 281.

177   The medium's penetration: Matthew Frye Jacobson and Gaspar González, *What Have They Built You to Do? The Manchurian Candidate and Cold War America* (Minneapolis: University of Minnesota Press, 2006), 96–97.

178   "citizenship was supplanted by viewership": Crary, *24/7*, 79.

178   treated like Pavlov's dog: Packard, *The Hidden Persuaders*, 32.

178   some Western social critics: Susan L. Carruthers, *Cold War Captives: Imprisonment,*

Escape, and Brainwashing (Berkeley: University of California Press, 2009), 231; J. Hoberman, *An Army of Phantoms: American Movies and the Making of the Cold War* (New York: The New Press, 2011), 308, 315.

178   "magic room of Communist propaganda": Lajos Ruff, *The Brain-Washing Machine* (London: Robert Hale, 1959 [1958]), 91.

178   "What possible immunity": Marshall McLuhan, *Understanding Media: The Extensions of Man* (New York: Mentor, 1964), 286.

179   eclipsed by the paranoia: Nadel, "Television"; Jacobson and González, *What Have They Built You to Do?*

179   Richard Condon's 1959 novel: Richard Condon, *The Manchurian Candidate* (New York: Four Walls Eight Windows, 2003 [1959]).

179   Richard Condon wrote: Richard Condon, "'Manchurian Candidate' in Dallas," *Nation*, December 28, 1963, 449–51.

180   Condon's novel treated: Condon, *The Manchurian Candidate*, 172–73, 185.

180   "Each patrol member": Ibid., 55.

181   "The first thing": Ibid., 32.

182   In the haunted dreams: Andreas Killen and Stefan Andriopoulos, "Editors' Introduction: On Brainwashing: Mind Control, Media, and Warfare," *Grey Room* 45 (2011): 7–17.

182   by aligning television: Jacobson and González, *What Have They Built You to Do?*, 84–89.

182   Raymond quickly turns up: Ibid., 130.

183   For an incisive analysis of the screen within a screen effect in this scene, see Stefan Andriopoulos, "The Sleeper Effect: Hypnotism, Mind Control, Terrorism," *Grey Room* 45 (2011): 97.

185   "trigger films": Michael Rogin, *Ronald Reagan The Movie: And Other Episodes in Political Demonology* (Berkeley, CA: University of California Press, 1987), 254; Timothy Melley, *The Covert Sphere: Secrecy, Fiction, and the National Security State* (Ithaca, NY: Cornell University Press, 2012), 147.

185   hot and cool media: McLuhan, *Understanding Media*, 44–45; Canales, "'A Number of Scenes in a Badly Cut Film,'" 250.

Chapter 6: Brain Science in the Visionary Mode

187   what exactly did it mean: Gordon M. Shepherd, *Creating Modern Neuroscience: The Revolutionary 1950s* (New York: Oxford University Press, 2010), 45.

187   one who remained: Thanks to Franny Nudelman for helping me understand this point.

188   flicker and sensory deprivation: Martin Lee and Bruce Shlain. *Acid Dreams: The CIA, LSD, and the Sixties Rebellion* (New York: Grove Press, 1985); Nicolas Langlitz, *Neuropsychedelia: The Revival of Hallucinogen Research Since the Decade of the Brain* (Berkeley: University of California Press, 2012); Michael Pollan, *How to Change your Mind: What The New Science of Psychedelics Teaches Us About Consciousness, Dying, Addiction, Depression, and Transcendence* (New York: Penguin, 2019), 207.

188   brain death: John Modern, "EEG," Somatosphere, January 12, 2014, http://somato

sphere.net/author/john-modern/; Christof Koch, "The Footprints of Consciousness," *Scientific American Mind* 28, no. 2 (March 2017): 52–59; Sean Quinlan, "Shots to the Mind: Violence, the Brain, and Biomedicine in Popular Novels and Film in Post-1960s America," *European Journal of American Culture* 32, no. 3 (2013): 215–34.

189   science fiction story "The Brain": Norbert Wiener, "The Brain," *Technical Engineering News*, April 1952; W. Katherine Hayles, *How We Became Posthuman: Virtual Bodies in Cybernetics, Literature, and Informatics* (Chicago: University of Chicago Press, 1999), 116.

189   Wiener's "electronic brain": Marshall McLuhan, *The Mechanical Bride: Folklore of Industrial Man* (Boston: Beacon Press, 1951); Norbert Wiener, *The Human Use of Human Beings* (New York: Da Capo, 1950), 129–30.

190   "whirling phantasmagoria": McLuhan, *The Mechanical Bride*, v; see also Charles R. Acland, *Swift Viewing: The Popular Life of Subliminal Influence* (Durham NC: Duke University Press 2012), 38–39.

190   "whirling spiral": W. Grey Walter, *The Living Brain* (London: Gerald Duckworth and Co., 1953), 112.

190   "We are but whirlpools": Wiener, *The Human Use of Human Beings*, 96.

191   he charted much: John Geiger, *Chapel of Extreme Experience: A Short History of Stroboscopic Light and the Dream Machine* (Brooklyn, NY: Soft Skull Press, 2003); Hayward, "The Tortoise and the Love-Machine"; Pickering, *The Cybernetic Brain*. See also Ferdia, "Should We Take an Anxiety Pill?—The Brains Trust," YouTube, May 3, 2019, https://www.youtube.com/watch?v=R-rrQUKAjiQ.

191   His novel: W. Grey Walter, *The Curve of the Snowflake* (New York: W. W. Norton, 1956), 137–38.

191   In the preface: Walter, *The Living Brain*, 11–12.

192   "enlargement of the field": Ellen Herman, *The Romance of American Psychology: Political Culture in the Age of Experts* (Berkeley: University of California Press, 1995).

192   geopolitical tensions of the Cold War: Walter, *The Living Brain*, xxi.

193   social organ: W. Grey Walter, "The Social Organ," *Impact of Science on Society* 18, no. 3 (1968): 179–86.

193   as one scholar puts it: Geiger, *Chapel of Extreme Experience*; Pickering, *The Cybernetic Brain*, 11–12.

194   In his correspondence: William S. Burroughs, *Rub Out the Words: The Letters of William S. Burroughs, 1959–1974*, edited by Bill Morgan (New York: Ecco, 2012), 58, 286, 314.

194   Writing to Walter: Ibid., 52.

194   flicker appeared to break down: Daniel Odier, *The Job: Interviews with William S. Burroughs* (New York: Penguin, 1974), 132.

194   In a letter: Burroughs to Ginsberg, November 2, 1960, in Burroughs, *Rub Out the Words*, 58.

195   "needle action": Burroughs to Joe Gross, October 17, 1968, in Burroughs, *Rub Out the Words*, 286.

195   he had become deeply suspicious: Burroughs to Charles Upton, January 2, 1970, in Burroughs, *Rub Out the Words*, 318.

195   In one letter: Burroughs to Anthony Balch, December 13, 1965, in Burroughs, *Rub Out the Words*, 192. See also William S. Burroughs, *Naked Scientology* (Bonn, Germany: Expanded Media Editions, 2000).

195   Burroughs discussed: Burroughs to Brion Gysin, March 13, 1969, in Burroughs, *Rub Out the Words*, 296.

195   "I have made": Barry Miles, *The Beat Hotel: Ginsberg, Burroughs and Corso in Paris, 1957–1963* (New York: Grove Press, 2000), 220.

196   looked at with the eyes closed: For a recent reappearance of the Dream Machine, see "Dreamachine Review—As Close to State-Funded Psychedelic Drugs as You Can Get," *Guardian*, May 9, 2022, https://www.theguardian.com/artanddesign/2022/may/09/dreamachine-review-as-close-to-state-funded-psychedelic-drugs-as-you-can-get.

196   As Burroughs subsequently related: Odier, *The Job*, 172–73. See also Branden W. Joseph, *Beyond the Dream Syndicate: Tony Conrad and the Arts After Cage* (New York: Zone, 2011), 307.

197   As the novelist John Clellon Holmes: John Clellon Holmes, "The Philosophy of the Beat Generation," *Esquire*, February 1, 1958, 33–36.

198   had experimented with yage: William S. Burroughs, *Queer* (New York: Viking, 1985), 116.

198   "brainwashing, thought control": Burroughs to Ginsberg, October 1956, quoted in David Seed, *Brainwashing: The Fictions of Mind Control* (Kent, OH: Kent State University Press, 2004), 134.

199   Dr. Benway: William S. Burroughs, *Naked Lunch* (New York: Grove Press, 1959), 25. See also Seed, *Brainwashing*, 136.

199   "biocontrol": Burroughs, *Naked Lunch*, 136.

199   *The Job* contains: Odier, *The Job*, 37–38, 40–47, 59; William S. Burroughs, *The Ticket That Exploded* (New York: Olympia, 1967 [1962]), 83; Burroughs, *Naked Scientology*.

200   "The will of the subject": Walter, *The Living Brain*, 106.

200   "mass deconditioning": Odier, *The Job*, 47.

200   In another such trial: Hayward, "The Tortoise and the Love-Machine," 628.

201   Burroughs's long-standing interest in medicine: A. J. Lees, *Mentored by a Madman: The William Burroughs Experiment* (Widworthy Barton, UK: Notting Hill, 2016).

201   he immersed himself: *Bio-feedback Newsletter*, April 1972, Folio 155, Box 60, William S. Burroughs Papers, Manuscripts and Archives Division, New York Public Library (hereafter NYPL).

201   The method's origins: "Biofeedback Training," Box 60, Folio 155, William S. Burroughs Papers, NYPL.

201   Electronic signals: Pickering, *The Cybernetic Brain*, 83 ff.

201   "pictures and films": Burroughs, *Naked Scientology*, 37.

202   "You don't have to tell me": Burroughs to *Los Angeles Free Press*, April 21, 1970, in Burroughs, *Rub Out the Words*, 331.

202   Dialectics of Liberation congress: Seed, *Brainwashing*, 134; Michael E. Staub,

Madness Is Civilization: When the Diagnosis Was Social, 1948–1980 (Chicago: University of Chicago Press, 2011), 118, 126.

202 "brainless": Pickering, *The Cybernetic Brain*, 172.

202 his lifelong experience of addiction: Seed, *Brainwashing*, 134–56.

203 a short piece: William S. Burroughs, "Switch On and Be Your Own Hero," *Mayfair*, June 1968, 52–54.

203 They envisioned the mass production: Geiger, *Chapel of Extreme Experience*, 66.

203 An unpublished piece: Brion Gysin and William Burroughs, "The Golden Dreamachine" and "Operation Sense Withdrawal," Box 10, folder 11, Folio 3, William S. Burroughs Papers, NYPL.

204 Gysin linked flicker: Geiger, *Chapel of Extreme Experience*, 55.

204 Gysin likened the dreamlike visions: Miles, *The Beat Hotel*, 221–22.

205 strobe became part: Fred Turner, *From Counterculture to Cyberculture: Stewart Brand, the Whole Earth Network, and the Rise of Digital Utopianism* (Chicago: University of Chicago Press, 2006).

205 short film "The Flicker": Geiger, *Chapel of Extreme Experience*, 73–77; Joseph, *Beyond the Dream Syndicate*.

Chapter 7: Brain Science Between Cold War and Counterculture

207 "the feeling little Alice had": A. R. Luria, *The Mind of a Mnemonist: A Little Book About a Vast Memory*, translated by Lynn Solotaroff (Cambridge, MA: Harvard University Press, 1987 [1967]), 73.

208 the writings of Darwin and Freud: John C. Lilly, *The Scientist: A Metaphysical Autobiography* (Oakland CA: Ronin Publishing, 1997 [1978]), 17.

209 "explorable brain": Rachel Elder, "Speaking Secrets: Epilepsy, Neurosurgery, and Patient Testimony in the Age of the Explorable Brain, 1934–1960," *Bulletin of the History of Medicine* 89, no. 4 (Winter 2015): 761–89.

209 his world was split: Lilly, *The Scientist*, 86–87.

210 Lilly's highly appreciative comments: John Lilly, "Comment," *Psychoanalytic Quarterly* 22, 1 (1953): 62–63.

210 In a letter: Lilly to Morison, October 17, 1951; Morison to Lilly, October 26, 1951; John C. Lilly, "Reciprocity Proposal: Relation Between Brain, Body, Mind," June 7, 1951, Box 31, John C. Lilly Papers.

211 "moving pictures": Morison, diary, April 15, 1949, RF, RAC.

211 "goes on its merry way": Lilly to Morison, October 26, 1951, Box 29, John C. Lilly Papers.

212 "So you want to hook": Lilly, *The Scientist*, 79–80.

212 Lilly described himself: Ibid., 83.

212 an "injured" brain: Lilly to Morison, October 5, 1951, Box 29, John C. Lilly Papers.

213 "John Lilly has been": Morison, diary, October 4, 1954, RF, RG 12, RAC.

213 without undue anxiety: Ibid.

214 He speculated: John C. Lilly, in *Illustrative Strategies for Research on Psychopathology in Mental Health*, Symposium no. 2 (New York: Group for the Advancement of Psychiatry, New York, June 1956), 18.

214   "natural rhythms": Lilly, *The Scientist*, 98–103.

214   "This morning we are considering": John C. Lilly, opening remarks, in *Methods of Forceful Indoctrination: Observations and Interviews*, Symposium no. 4 (New York: Group for the Advancement of Psychiatry, July 1957), 233–34.

215   "human recording and transmitting apparatus": Robert J. Lifton, "Psychiatric Aspects of Chinese Communist Thought Reform," in *Methods of Forceful Indoctrination: Observations and Interviews*, 245.

215   "I believe that the literature": Lilly, in *Methods of Forceful Indoctrination*, 293.

216   "a potential future": James Miller, in *Methods of Forceful Indoctrination*, 295.

216   "will not be imposed": Lilly, "LSD—Reward and Punishment," n.d., Box 28, John C. Lilly Papers.

217   "I myself experience": Lilly to Kubie, September 28, 1961, Kubie Correspondence, John C. Lilly Papers.

217   "experimental intervention": Cathryn B. A. Walters, Jay T. Shurley, and Oscar A. Parsons, "Differences in Male and Female Responses to Underwater Sensory Deprivation: An Exploratory Study," in *Journal of Nervous and Mental Disease* 135, no. 4 (1962): 302–10.

218   "perception without object": Woodburn Heron, "Cognitive and Physiological Effects of Perceptual Isolation," in *Sensory Deprivation: A Symposium Held at Harvard Medical School*, edited by Philip Solomon et al. (Cambridge, MA: Harvard University Press, 1961), 17.

218   "Lilly's experimental amniotic fluid": Lawrence Kubie, "Theoretical Aspects of Sensory Deprivation," in Solomon, *Sensory Deprivation*, 212, 215.

218   "piper of whale-hugging": D. Graham Burnett, "A Mind in the Water," *Orion*, May–June 2010, 14.

218   Lilly evidently also understood: Lilly, *The Scientist*, 29, 105.

219   Lilly addressed: John C. Lilly and Jay T. Shurley, "Experiments in Solitude, in Maximum Achievable Physical Isolation with Water Suspension, of Intact Healthy Persons." First given at the 1958 Harvard Medical School Symposium on Sensory Deprivation, this paper was later published in Bernard E. Flaherty, ed., *Psychophysiological Aspects of Space Flight* (New York: Columbia University Press, 1961), 238–47. See also "Experiments in Solitude," 5A1, Box 5, John C. Lilly Papers.

219   the Russians might already have surpassed: Lilly, "Human Brain Research and Ethics," ca. 1960, 5A1, Box 5, John C. Lilly Papers. See also Lilly, "No Holds Barred," 5A1, Box 5, John C. Lilly Papers.

219   He went on to explain: Lilly and Shurley, "Experiments in Solitude." See also Lilly, "No Holds Barred."

220   "I deeply feel": Lilly to Kubie, December 13, 1967, Kubie Correspondence, John C. Lilly Papers.

220   he became convinced: John C. Lilly, *Programming and Metaprogramming in The Human Biocomputer: Theory and Experiments* (New York: Julian Press, 1968), 7.

220   By contrast, Penfield: Wilder Penfield, *Mystery of the Mind: A Critical Study of Consciousness and the Human Brain* (Princeton, NJ: Princeton University Press, 1975), 79.

220   one exchange of letters: Lilly to Hebb, May 19, 1956; Hebb to Lilly, June 8, 1956, Box 2a, John C. Lilly Papers.

220   an unpublished document: John C. Lilly, "Special Considerations of Modified Human Agents as Reconnaissance and Intelligence Devices," 5A1, Box 5, John C. Lilly Papers. Indications in his not fully catalogued papers at Stanford are that Lilly first delivered this paper at the GAP symposium on coercive indoctrination in 1956, where he took such pains to dismiss Orwell's account of brainwashing. Appended to it are Lilly's handwritten notes on that symposium, in which he describes the existing state of knowledge on the subject as primitive and the results, from an operational standpoint, "practically zero." More research, he stated, was urgently needed.

221   "psychic implants": D. Ewen Cameron, "Psychic Driving," *American Journal of Psychiatry* 112, no. 7 (1956): 502–9. See also Cameron, "Sensory Deprivation," in Flaherty, *Psychophysiological Aspects of Space Flight*, 225–37.

221   He noted, for instance: Lilly, "Special Considerations of Modified Human Agents as Reconnaissance and Intelligence Devices," 4.

221   In another unpublished text: John C. Lilly, "Human Brain Research and Ethics," 5A1, Box 5, John C. Lilly Papers.

222   "Is the brain": Lilly, *The Scientist*, 85.

223   "mystical character": Lilly, in *Illustrative Strategies for Research on Psychopathology in Mental Health*, 43.

223   contains an avowedly mystical component: Andrew Pickering, *The Cybernetic Brain: Sketches of Another Future* (Chicago: University of Chicago Press, 2010), 206.

223   "our universe and another one": John C. Lilly, *The Center of the Cyclone: An Autobiography of Inner Space* (New York: Julian Press, 1972), 53.

223   Lilly's reference: Lilly, *The Scientist*, 170–73.

224   bears many traces: Charlie Williams, "On 'Modified Human Agents': John Lilly and the Paranoid Style in American Neuroscience," *History of the Human Sciences* 32, no. 5 (2019): 84–107. Williams's discussion of Lilly is particularly good.

224   In those scenes: Cathy Gere, "'Nature's Experiment': Epilepsy, Brain Localization, and the Emergence of the Cerebral Subject," in *Neurocultures: Glimpses into an Expanding Universe*, edited by Francisco Ortega and Fernando Vidal (Frankfurt am Main, Germany: Peter Lang, 2011). I thank Patrick Kilian for his illuminating insights into this topic.

224   To recap this remarkable itinerary: Lilly, *The Scientist*, 76.

226   migraine was a constant: Lilly, *The Center of the Cyclone*, 51–55; Lilly, *The Scientist*, 29, 144–45.

226   the kinship among hallucinations: Oliver Sacks, *Migraine* (New York: Vintage, 1992), 196.

226   his late book: Lilly, *Programming and Metaprogramming*.

226   first drafted in the early 1950s: John C. Lilly, "A Proposal for a Research Program on the Relations Between the Activities of the Brain, Body and Mind," June 7, 1951–May 15, 1952, Box 31, John C. Lilly Papers.

227   a childhood trauma: Lilly, *The Scientist*, 41.

227   "natural experiment": Lilly, "Proposal for a Research Program."

Conclusion

228   "in jars and on slides": "Soviet Turmoil; Moscow Saving Great Brains," *New York Times*, September 9, 1991.

229   Eventually the brains: Paul Gregory, *Lenin's Brain and Other Tales from the Secret Soviet Archives* (Stanford, CA: Hoover Institution Press, 2008).

229   Postmortem investigation revealed: Andrew Osborn, "Joseph Stalin 'Had Degenerative Brain Condition,'" *Telegraph*, April 21, 2011; see also "Kremlin Case History," *Time*, March 16, 1953.

229   Among the documents included: Gregory, *Lenin's Brain and Other Tales from the Secret Soviet Archives*, 34.

229   Was it a coincidence: Christof Koch, "The Footprints of Consciousness," *Scientific American Mind* 28, no. 2 (March 2017): 52–59; Jan Plamper, *The History of Emotions: An Introduction* (New York: Oxford University Press, 2015), 61.

230   "the deeper meanings": John Brockman, cited in Plamper, *The History of Emotions*, 224.

230   The putative universalism: Plamper, *The History of Emotions*, 240; Suparna Choudhury and Jan Slaby, "Introduction," in *Critical Neuroscience: A Handbook of the Social and Cultural Contexts of Neuroscience*, edited by Suparna Choudhury and Jan Slaby (London: Wiley-Blackwell, 2012); Curtis White, *The Science Delusion: Asking the Big Questions in a Culture of Easy Answers* (Brooklyn, NY: Melville House, 2013).

230   physics ceded to the life sciences: Ian Hacking, introductory essay, in Thomas S. Kuhn, *The Structure of Scientific Revolutions*, 50th anniversary edition (Chicago: University of Chicago Press, 2012 [1962]), viii; Plamper, *The History of Emotions*, 61.

230   The iconic image: Sally Satel and Scott O. Lilienfeld, *Brainwashed: The Seductive Appeal of Mindless Neuroscience* (New York: Basic Books, 2013), x.

230   Mind, declared Eric Kandel: Eric R. Kandel, *In Search of Memory: The Emergence of a New Science of Mind* (New York: W. W. Norton, 2006), xii.

230   Biology was similarly recast: Nicolas Langlitz, *Neuropsychedelia: The Revival of Hallucinogen Research Since the Decade of the Brain* (Berkeley: University of California Press, 2012), 2; Nikolas Rose, *The Politics of Life Itself: Biomedicine, Power, and Subjectivity in the Twenty-First Century* (Princeton, NJ: Princeton Univerity Press, 2007).

230   The twentieth-century view: Cornelius Borck, *Brainwaves: A Cultural History of Electroencephalography* (New York: Routledge, 2018), 8.

230   "invested with ever more extravagant hopes": Katja Guenther, *Localization and Its Discontents: A Genealogy of Psychoanalysis and the Neuro Disciplines* (Chicago: University of Chicago Press, 2015), 8.

231   was transferred to scientists: Plamper, *The History of Emotions*, 225.

231   Is it that neuroscience offers: Satel and Lilienfeld, *Brainwashed*, xiv; Plamper, *The History of Emotions*, 61, 240.

231   have been challenged: Satel and Lilienfeld, *Brainwashed*, 12–14; Plamper, *The History of Emotions*, 230.

231   biopsychiatry has failed: Anne Harrington, *Mind Fixers: Psychiatry's Troubled Search*

for the Biology of Mental Illness (New York: W. W. Norton, 2019); Satel and Lilienfeld, *Brainwashed*, 23.

231 "brain over-claim syndrome": Morse, quoted in Satel and Lilienfeld, *Brainwashed*, 160, n. 25.

231 "cosmetic neurology": Peter Kramer, *Listening to Prozac: A Psychiatrist Explores Antidepressant Drugs and the Remaking of the Self* (New York: Viking, 1993); Margaret Talbot, "Brain Gain: The Underground World of 'Neuroenhancing' Drugs," *New Yorker*, April 27, 2009.

231 President's Council on Bioethics: *Beyond Therapy: Biotechnology and the Pursuit of Happiness* (Washington, DC: The President's Council on Bioethics, 2003). See also Aldous Huxley, new foreword to *Brave New World* (New York: Penguin, 1946).

232 optimizing performance: Steven Johnson, *Mind Wide Open: Your Brain and the Neuroscience of Everyday Life* (New York: Scribner, 2004), 72–75; Borck, *Brainwaves*, 264–65; Andrew Pickering, *The Cybernetic Brain: Sketches of Another Future* (Chicago: University of Chicago Press, 2010), 83.

232 "free societies": *Beyond Therapy*, 302.

232 it also seems true: Jia Tolentino, *Trick Mirror: Reflections on Self-Delusion* (New York: Random House, 2019), 12.

232 Toward the end of the story: Adam Johnson, *The Orphan Master's Son* (New York: Random House, 2012), 316–17.

233 A veil was gradually pulled back: Jane Kramer, *The Dark Side: The Inside Story of How the War on Terror Turned into a War on American Ideals* (New York: Anchor, 2009).

233 *KUBARK*: Alfred McCoy, *A Question of Torture: CIA Interrogation, from the Cold War to the War on Terror* (New York: Holt, 2006); Jonathan D. Moreno, *Mind Wars. Brain Research and National Defense* (New York: Dana Press, 2006).

233 relationships among secrecy, democracy, and torture: Darius Rejali, *Torture and Democracy* (Princeton, NJ: Princeton University Press, 2007), 458.

233 techniques of over- and understimulation: Naomi Klein, *The Shock Doctrine: The Rise of Disaster Capitalism* (New York: Picador, 2008); see also Charlie Savage, "'Guantánamo Diary' Writer Is Sent Home to Mauritania," *New York Times*, October 17, 2016, https://www.nytimes.com/2016/10/18/us/guantnamo-diary-writer-mohamedou-ould-slahi.html.

233 Such revelations serve also: Jane Mayer, *The Dark Side: The Inside Story of How the War on Terror Turned into a War on American Ideals* (New York: Anchor, 2009); McCoy, *A Question of Torture*.

233 Martin Seligman: "Report of the Independent Reviewer and Related Materials," American Psychological Association, 2015, https://www.apa.org/independent-review, 48–49, 126.

234 pure ideology: White, *The Science Delusion*, 10.

234 "Freudianism and Marxism": Tom Wolfe, "Sorry, but Your Soul Just Died," *Forbes*, December 2, 1996.

235 the brain in a vat: Antonio Damasio, *Descartes' Error: Emotion, Reason and the Human Brain* (New York: Harcourt, 1994), 250.

235   "such a brain": Ibid., 227–28.

235   This has resulted: Ibid., 255.

235   cast some doubt: Ian Hacking, "Our Neo-Cartesian Bodies in Parts," *Critical Inquiry* 34, no. 1 (2007): 78–105; Langlitz, *Neuropsychedelia*, 209; Plamper, *The History of Emotions*, 17–19.

235   "a bunch of neurons": Siri Hustvedt, *A Woman Looking at Men Looking at Women: Essays on Art, Sex, and the Mind* (New York: Simon & Schuster, 2016), 89.

235   a new biochemical model: David Healy, *The Creation of Psychopharmacology* (Cambridge, MA: Harvard University Press, 2002), 355; Rose, *The Politics of Life Itself*.

235   a gradual flattening out: Langlitz, *Neuropsychedelia*, 2; Rose, *The Politics of Life Itself*.

236   This is true equally: Edward Shorter, *A History of Psychiatry: From the Era of the Asylum to the Age of Prozac* (New York: Wiley, 1998).

236   To some critics: Allen Frances, "Entrenched Reductionisms: The Bête Noire of Psychiatry," *History of Psychology* 19, no. 1 (2016): 57–59; Michael Pollan, "My Adventures with the Trip Doctors," *New York Times*, May 15, 2018, https://www.nytimes.com/interactive/2018/05/15/magazine/health-issue-my-adventures-with-hallucinogenic-drugs-medicine.html.

236   efforts to forge such a synthesis persist: Guenther, *Localization and Its Discontents*, 189.

236   The deactivation of inhibiting mechanisms: Langlitz, *Neuropsychedelia*; Michael Pollan, *How to Change your Mind: What the New Science of Psychedelics Teaches Us About Consciousness, Dying, Addiction, Depression, and Transcendence* (New York: Penguin Press, 2019), 317.

237   Though the ethical issues raised: Gordon M. Shepherd, *Creating Modern Neuroscience: The Revolutionary 1950s* (New York: Oxford University Press, 2010), 172.

237   some authors implicated Hebb: Klein, *Shock Doctrine*; McCoy, *A Question of Torture*; but see Richard E. Brown, "Alfred McCoy, Hebb, the CIA and Torture," *Journal of the History of the Behavioral Sciences* 43, no. 2 (2007). See also *Opinions of George Cooper, Q.C., Regarding Canadian Government Funding of the Allan Memorial Institute in the 1950's and 1960's* (Ottawa, Ontario, Canada: Minister of Supply and Services Canada, 1986), https://archive.org/details/GeorgeCooperReportEwenCameron1986. Naomi Klein's website contains links to exchanges between Hebb and the Canadian Defense Council in the early 1950s about this work; see https://tsd.naomiklein.org/shock-doctrine/resources/part1/chapter1.html.

237   "fundamental disturbance": Harold Wolff, SIHE Annual Report, 1957, Box 6, folder 15, Harold Wolff Papers.

238   "motion-picture like hallucinations": D. O. Hebb, "The Mind's Eye," *Psychology Today*, May 1969. See also Stanley Cobb, foreword, in *Sensory Deprivation: A Symposium Held at Harvard Medical School*, edited by Philip Solomon et al. (Cambridge, MA: Harvard University Press, 1961), xvii.

238   Just as the Cold War served: Moreno, *Mind Wars*.

238   Should memory be off limits: Joseph LeDoux, *Anxious: Using the Brain to Understand and Treat Fear and Anxiety* (New York: Penguin Books, 2015), 283.

238   in order to prevent the encoding: *Beyond Therapy*, 226.

238   "Nor can we forget": Ibid., 233. Rose suggests that the worries of the authors of this document are exaggerated; that authors such as Peter Kramer do not claim that the mind can be reshaped by pills alone. See Rose, *The Politics of Life Itself*, 98.

239   the continuing opacity of these images: White, *The Science Delusion*, 212; Danielle Carr, "Brain Scans Look Stunning, but What Do They Actually Mean?," Psyche, November 22, 2021, https://psyche.co/ideas/brain-scans-look-stunning-but-what-do-they-actually-mean.

239   "Together with little Alice": A. R. Luria, *The Mind of a Mnemonist: A Little Book About a Vast Memory*, translated by Lynn Solotaroff (Cambridge, MA: Harvard University Press, 1987 [1967]), 73.

240   "If I want something": Ibid., 138–39.

240   For instance, he achieved: Ibid., 142.

240   even when he acknowledged: Oliver Sacks, *Hallucinations* (New York: Vintage, 2012), 154–55.

240   the damaged brain's resilience: Oliver Sacks, *On the Move: A Life* (New York: Vintage, 2014), 270.

241   the "cybernetic" model: Sacks, *Hallucinations*, 147.

241   "neurosurgery of identity": Oliver Sacks, *The Man Who Mistook His Wife for a Hat* (New York: Simon & Schuster, 1970), 147–48.

241   "analysis in depth": Ibid., 141.

241   Some of Mrs. O'C.'s reminiscences: Ibid., 144–45.

242   Just how resilient: Langlitz, *Neuropsychedelia*, 262 ff; Rose, *The Politics of Life Itself*, 98–99.

242   NeuroSky MindWave: Jasper Craven, "America's Legion," *The Baffler*, November 2019, 123; Moreno, *Mind Wars*.

242   the brains depicted: White, *The Science Delusion*, 169.

242   anxious reflections of Vance Packard: Satel and Lilienfeld, *Brainwashed*, 25–26.

242   ideology of capitalism: White, *The Science Delusion*, 11.

242   "It used to be": Eva Galperin, quoted in Sharon Weinberger, "Private Surveillance is a Lethal Weapon Anybody Can Buy," *New York Times*, July 19, 2019, https://www.nytimes.com/2019/07/19/opinion/private-surveillance-industry.html.

243   a thoroughly cultural object: Max Stadler, "The Neuromance of Cerebral History," in Choudhury and Slaby, *Critical Neuroscience*, 137.

Index

Page numbers in *italics* refer to illustrations.

About the Author

Andreas Killen is a professor of history at City College and the CUNY Graduate Center. He is the author of *1973 Nervous Breakdown*, *Berlin Electropolis*, and *Homo Cinematicus*. His writing has appeared in Salon and the *New York Times Magazine*.